Waste in Textile and Leather Sectors

Edited by Ayşegül Körlü

Published in London, United Kingdom

IntechOpen

Supporting open minds since 2005

Waste in Textile and Leather Sectors
http://dx.doi.org/10.5772/intechopen.90014
Edited by Ayşegül Körlü

Contributors
Burcu Okutucu, Patricia Vázquez, Katerine Igal, Burçin Ütebay, Ahmet Çay, Pınar Çelik, Caciora Tudor, Ilieş Dorina Camelia, Herman Grigore Vasile, Ilieş Alexandru, Indrie Liliana, Wendt Jan, Axinte Anamaria, Diombera Mamadou, Berdenov Zharas, Lite Cristina, Candan Akca, Kgomotso N.G. Mtshatsheni, Augustine E. Ofomaja, Bobby E. Naidoo, Ali Yorgancioglu, Bahri Başaran, Aykut Sancakli, Cevza Candan, Sedef Uncu Akı, Banu Nergis, Neslihan Sebla Önder, Nilgün Özdil, Gonca Özçelik Kayseri, Gamze Süpüren Mengüç, Lorena Henrique Guimarães Coelho, Catarina Costa, Manuela Pintado, Patrícia Batista, Bruna Moura, Ana Isabel Magalhães, Pedro Faria, Rosa Maria, Albertina Reis, Marta Carvalho, Mário Marques, Ângela Teles, Regina Malgueiro, Sara Fernandes, José De Almeida Morgado, Augusta Marinho, Maria Helena Vilaça, Jéssica Alexandra Pereira, Pedro José Magalhães, Ana Sofia Silva, Ricardo Jorge Silva, Mario Jorge Silva, Vera Lúcia Sá, Sandra Gabriela Ventura, João Silva Abreu, Joaquim Manuel Gaião, Raquel Rosa Mourão, Fernando Manuel Merino, Mónica Sofia Gonçalves, Albu Adina

First published in London, United Kingdom, 2020 by IntechOpen
IntechOpen is the global imprint of INTECHOPEN LIMITED, registered in England and Wales, registration number: 11086078, 7th floor, 10 Lower Thames Street, London, EC3R 6AF, United Kingdom
Printed in Croatia

British Library Cataloguing-in-Publication Data
A catalogue record for this book is available from the British Library

Additional hard and PDF copies can be obtained from orders@intechopen.com

Waste in Textile and Leather Sectors
Edited by Ayşegül Körlü
p. cm.
Print ISBN 978-1-78985-243-1
Online ISBN 978-1-78985-244-8
eBook (PDF) ISBN 978-1-83880-497-8

For EU product safety concerns:
IN TECH d.o.o., Prolaz Marije Krucifikse Kozulić 3, 51000 Rijeka, Croatia, info@intechopen.com or visit our website at intechopen.com.

Meet the editor

Prof. Dr. Ayşegül Körlü received her Msc and PhD degrees in textile engineering from Ege University. She has been employed by the Department of Textile Engineering, Engineering Faculty, Ege University since 1988. Currently, her main research interests are sustainable and ecological finishing processes, pretreatment of natural fibers, functional textiles and occupational safety in textile industry. She has co-authored numerous publications and has taken part in the management of many national and international projects.

Contents

Preface

Changing lifestyles and living conditions has caused increases in industrial and domestic waste since the industrial revolution. Therefore, human and environmental health is one of the most critical problems humanity faces today. Industrial waste from the textile and leather industry can be studied under four classes: air, water, solid waste pollution, and noise. Textile and leather products are part of our daily life. If they fall into disfavor, textile and leather products will be a domestic waste. Waste management for solving environmental problems is realized by reduction of waste, reuse, material recycling, material and energy recovery, storage without recycling, or disposal by burning.

In this book, the relationship between the textile/leather materials and waste is examined from many viewpoints. The book consists of ten chapters.

The first chapter is about antimicrobial finishing. The objective of this work is to synthesize modified silica by including Ag and C and then adding antimicrobial additives to obtain antimicrobial fabrics. Disused batteries are used as the C source. A waste product that does not originate from the textile industry is evaluated using antimicrobial textile finishing. The authors suggest that the method is applied to the treatment of waste with high metal contents such as batteries, computer waste, slags from the metallurgical industry, etc.

The second chapter is about the waste problem around antimicrobial finishing. Antimicrobial agents ensure hygienic conditions and prevent the spread of diseases, especially in crowded places such as hospitals, baby nurseries, and barracks. On the other hand, antimicrobial finishing has adverse effects such as being toxic, allergy-causing, and carcinogenic on the wearer and all the living organisms during the antimicrobial textile product's production and serving life. In the second chapter, the author mainly discusses metal-based antimicrobial finishing and triclosan-based antimicrobial finishing.

The third chapter is about textile waste recycling. The authors present a general evaluation of the management of textile wastes in terms of ensuring sustainability and minimizing environmental impacts.

The fourth chapter is about denim recycling. The authors analyse different recycling technologies and focus on denim fabrics produced with recycled content.

The fifth chapter is about the recycling of textile waste and the use of them as a sound absorber. The authors compare conventional polyester and polypropylene nonwoven insulation materials with the nonwoven fabrics produced from recycled materials.

In the sixth chapter, the authors focus on the potential for reusing natural by-products from different sources such as the dairy industry, agriculture, and leather industry. In this context, they present a case study about the new generation of coated textile solutions that can be used as an alternative to natural and/or synthetic leather.

The seventh chapter is about value addition to the leather industry wastes and by-products. The authors focus on the evaluation of the tannery solid wastes and by-products by partial and total denaturation and hydrolyzation.

The eighth chapter is about the characterization of grafted acrylamide onto the pine magnetite composite for the removal of methylene blue from wastewater. The authors concentrate on chemical modification of natural biomass through grafting and use of them as an adsorbent for the treatment of wastewater.

In the ninth chapter, the author summarizes the specific components of molecularly imprinted polymer (MIP), synthesizing methods, and some examples of treatment of wastewater by MIPs.

The tenth chapter is about human health and historical textiles. The authors focus on determining the degree of conservation of the materials and their implications on the health of the people with whom they come into contact.

Briefly, the relationship between textiles and leather with the waste both as a source of pollution and as a solution is mentioned in this book.

I thank all the authors contributing to the book and hope that it will be helpful to the readers. The preparatory works of this book were made during the COVID-19 period. Therefore, I dedicate this book to healthcare providers fighting COVID-19 around the world.

Ayşegül Körlü
Engineering Faculty,
Department of Textile Engineering,
Ege University,
İzmir, Turkey

Chapter 1

Antimicrobial Fabrics Impregnated with Ag Particles Included in Silica Matrices

Katerine Igal and Patricia Vázquez

Abstract

A hospital that has a high incidence of acquired infections during the stay of patients in it is not considered efficient, since as Florence Nightingale, an English lady who died in 1910 and founder of the modern school of nursing, said: "the first thing that doesn't a hospital must do is get sick." Filamentous fungi, given their ability to grow on various substrates, are considered within the most damaging organisms. Among the fungi that are generally found in environments inhabited by humans in urban areas, we can mention Alternaria, Aspergillus, and Cladosporium, among others. With the incorporation of biocides into textiles, different methodologies are being studied depending on the stage they are performed; if it is at the finishing of the fabric, among the most used methods is the pad-dry-cure. The objective of this work was to synthesize modified silica by including Ag and C, where the latter is extracted from disused batteries and then added as antimicrobial additives to obtain antimicrobial fabrics.

Keywords: silica, silver, carbon, antimicrobial additives, antimicrobial fabrics

1. Introduction

The concern of human beings related to health care has always existed, and the increase in diseases caused by the enormous population density has forced us to look for effective technological solutions. Materials such as textiles, used for fabric production, can be easily colonized by a high amount of microorganism or can even be deteriorated by them. The microbial colonization on fabrics generates esthetic problems and can also lead to the degradation of the material, leaving it in disuse. Fungi are heterotrophic organisms that commonly colonize organic surfaces, such as coatings used in construction materials, paints, or fabrics, and due to the substrate type, they can be metabolized by them. This not only generates problems into the domestic environment, where many objects are built by organic substrates, but also the fungal growth can affect the human health by the production of allergens, irritants and mycotoxins. Therefore, antimicrobial additives need to be not only effective in fungal growth control but also safe and environmentally friendly chemical substances in their preparation.

In the last decades, different impregnation methods in fabrics [1, 2] and a wide variety of antimicrobial additives such as silver, quaternary ammonium salts [3], polyhexamethylene biguanide [4], triclosan [5], and chitosan [6], N-halamine compounds [7], and peroxy acids [8] have been studied. For example, an ecological and viable method has been used to re-coat cotton fabrics with silver nanoparticles [9, 10]. In studies conducted by Mahltig et al. [11], the sol-gel method was used to coat textiles with inorganic SiO_2 by the construction of layers containing Ag nanoparticles. The formation of these particles was investigated according to the curing treatment variables performed after the coating. Inorganic coatings containing Ag inhibited the growth of *Aspergillus niger* fungal strain and *Bacillus subtilis* and *Pseudomonas putida* bacteria [12].

Tomšič et al. [13] studied dry curing and the method was thoroughly compared. The antimicrobial solution was prepared from different concentrations of dispersed commercial silver chloride, with a reactive organic-inorganic binder (RB) using cotton fabrics. Washing cycles were carried out and then antifungal (against *Aspergillus niger* and *Chaetomium globosum*) and antibacterial assays (*Escherichia coli*) were evaluated, being more effective the exhaustive method, and also better results were obtained against bacteria compared with fungi. The results were different according to the Ag concentrations and the method of application in the cotton fabrics.

In other work [14], Ag nanoparticles have been used within polystyrene-block-polyacrylic acid copolymer (PS-b-PAA) micelle nuclei, synthesized by the free-radical polymerization method, in different relations. It has been determined that the impregnation method into the fabric is by an esterification reaction between PAA and the hydroxyl groups on the surface of the fabric. Another method studied is the use of new nanostructures and techniques that allow the production of nanoparticles for its application, in various sectors, in order to improve processes and increase productivity. For example, electrospinning method [15] is simple, inexpensive, and used in a wide variety of materials, making it one of the most used. The structures obtained have unique characteristics, such as large contact area and high porosity. Due to these properties, nanofibers are of great interest to be applied in different areas, such as biomedical, textile, and food, obtaining beneficial results.

Textiles and clothing are in daily contact with microorganisms of the environment and the human skin. In general, fabrics can be an excellent substrate for microbial growth, because they have an organic composition that provides an adequate base for human sweat and biofilm fixing. The human skin contains a complex mixture of microorganisms; even a "clean" skin has a typical population of between 100 and 1000 microorganism/cm^2. At these levels, they do not represent a health problem or bad smell. On the contrary, its presence and a balanced population are essential for human health [16]. But when the optimal growth conditions are provided, microorganisms are multiplied rapidly and can produce problems such as the generation of odors, loss of performance, discoloration of fabrics, and possibly infection. In the most extreme case, microorganisms can produce serious problems, such as rotting of the fabric, stains, unpleasant odors and health problems ranging from simple discomfort to physical irritation, allergic sensitization, toxic responses, infections, and diseases. Many of the characteristic malodors associated with the human body are due to the presence of large populations of microorganisms. Therefore, control of the undesirable effects of microbes on textiles is becoming an important issue in the textile industry. Currently, there is much interest in hygienic fabrics that offer an advantage compared with cleaning and odor prevention as a result of their antimicrobial properties or reduced bacterial growth.

The advantage of using an oxide matrix associated with an antimicrobial agent resides in its protective function that lead to a longer useful life of the material obtained maintaining its bioactivity over time, and the sol-gel method is an appropriated way to obtain this immobilization [17–19].

The aim of the present work was to obtain silica-based solids whose active phase is formed by C, from the recycling of disused batteries, and Ag particles.

In relation to the used carbon, the idea of recycling the metals that make up the batteries is to look at the exhausted batteries as a resource and not as a waste, since they contain valuable metals in high concentrations that, if the batteries are thrown in the trash, contaminate soil and water; on the other hand, if they are seen as raw material for the recovery of metals, they become an important resource since minerals are reused that are otherwise extracted from the mines with the consequent environmental impact that this activity entails.

To obtain recycled Zn, different elements are obtained, including coal, which is obtained by means of a biohydrometallurgical process for the treatment and recovery of metals from spent batteries. Biohydrometallurgy is a branch of metallurgy that uses biooxidation and/or bioleaching processes, referred to by a general term: biomining [20–24]. Biohydrometallurgy is the application of microbiological processes for the recovery of metals, mainly used by the mining industry. The acid bioleaching of unsaturated minerals uses the acidic reducing medium biogenerated by microorganisms for the solubilization of metals.

The materials synthesized were characterized by potentiometric titration, textural properties, X-ray diffraction (XRD), Fourier transform infrared spectroscopy (FTIR), scanning electron microscopy (SEM), and transmission electron microscopy (TEM). Subsequently, the synthesized solids were used as antimicrobial additives in fabrics, using the pad-dry-cure method. *Aspergillus* sp., *Chaetomium globosum*, and *Cladosporium* sp. fungi were selected to evaluate the antifungal activity from biodeteriorated fabrics [25]. These fungi were selected for their ability to grow on indoor surfaces and to be negative to human health to produce a variety of different compounds including mycotoxins [26–28]. The agar diffusion method was used to evaluate the antibacterial activity of impregnated fabrics (*E. coli* and *S. aureus*).

This work is focused on the preparation of new materials that can provide solutions to the technological and environmental challenges in different areas.

2. Materials and methods

2.1 Synthesis of the additives

The solids were synthesized by the sol-gel method, under N_2 atmosphere in order to reach an inert condition. Tetraethyl orthosilicate (TEOS) (Aldrich, 98%) was used as the precursor for silica solids, absolute ethanol (EtOH, Baker 99.9%) and distilled water as solvents, and ammonium hydroxide as catalyst (basic hydrolysis). The precursor/catalyst/EtOH/H_2O molar ratios were 1:1:5:4, respectively, in all the synthesis. Besides, different amounts of carbon, recycled from zinc-carbon batteries, were added into the reaction mixture.

Zinc-carbon batteries contain a cathode, which is a mixture of manganese oxide and conductive carbon, usually in the form of black [29] coal, and the anode, composed of a high-purity zinc alloy, is also the container that encloses battery active materials. The recycled carbon was milled with a ball mill, then sieved with a mesh of 200, and, finally, added to the synthesis. The obtained mixture was stirred for 2 h and dried at room temperature for 1 week. The nomenclature of samples are as follows: SB (without C), SB1 (0.1 C p/p), SB2 (1 C p/p), and SB3 (10 C p/p), respectively. Subsequently, two samples were selected: sample SB (without C), which was modified including during the synthesis process 4% w/w of silver nitrate (Aldrich, 99.9%), was called SBAg, and the sample SB3 (10% w/w C) which was treated in the same way including Ag and was called SB3Ag.

2.2 Characterization of the additives

The acidic properties of the solids were evaluated by potentiometric titration with *n*-butylamine, in a Metrohm 794 Basic Titrino titrator (Switzerland) with a double-junction electrode. First, 0.025 g of sample was suspended in 45 mL of acetonitrile and stirred for 540 s, and second, 0.025 mL/min of an *n*-butylamine solution in acetonitrile (0.025 N) was added, while stirring constantly. The textural properties of the additives, such as the specific surface area (S_{BET}), the pore volume, and pore size, were determined by adsorption/desorption in Micromeritics Accusorb 2100 equipment (USA), using N_2 as absorbable gas at 77 K. Before the measurement, each sample was degassed at 100°C for 12 h and under 30 mmHg. The X-ray diffraction (XRD) diagrams were obtained in Philips (Holland) PW-1390 (channel control) and PW-1394 (motor control) equipment coupled to a scanning graphical recorder, using Cu Kα (α = 1.5417 Å) radiation, Ni filter, 20 mA and 40 kV in the voltage source, a 5–60 2θ scanning angle range, a scanning rate of 2°/min, and 2000 counts/s for the amplitude of the vertical scale. Fourier transform infrared spectroscopy (FT-IR) spectra were obtained using Bruker Vertex 70 equipment (Germany) and pellets of the sample in KBr (Aldrich, 99 wt% FT-IR purity), measured in a range between 400 and 4000 cm^{-1} at room temperature. Two hundred scans were collected at a resolution of 4 cm^{-1} and averaged. Scanning electron microscopy (SEM) was used to obtain different micrographs of the additives, in Philips 505 equipment (Holland), using a voltage of 15 kV. Samples were supported on graphite and metallized with a sputtered gold film. The micrographs were obtained with an ADDAII acquisition device (Soft Imaging System). Transmission electron microscopy (TEM) was performed with a JEOL microscope (100 CX) (Japan), with an accelerating voltage of 100 kV. Samples were prepared by their suspension in ethanol and placing an aliquot over carbon-coated copper grids, allowing the samples to dry in a desiccator for 30 min at room temperature. X-ray mapping was acquired by using a Talos F200X HR-TEM microscope operating at 200 kV equipped with a SuperX EDS spectrometer (composed of 4 EDS SDD detectors).

2.3 Assessment of the antifungal activity of the additives

Aspergillus sp., *Chaetomium globosum*, and *Cladosporium* sp. fungi were selected to evaluate the antifungal activity of the solids, based on their cellulolytic ability in agar plate assays. *Aspergillus* sp. and *Cladosporium* sp. were previously isolated from biodeteriorated fabrics by conventional microbiological techniques, whereas *C. globosum* was selected from the CIDEPINT culture collection [30]. From subcultures growing in Petri dishes, inoculums of cited fungi were obtained using a solution of 0.85% p/v NaCl and 0.005% p/v Tween 20, being the concentration of the suspension adjusted to 106 spores/mL employing a Neubauer chamber. The composition of the culture medium used was 1.5 g agar (Parafarm), 1 g dextrose (Anedra, analytical reagent), 0.5 g proteose peptone (OXOID), 0.1 g KH_2PO_4 (Anedra, analytical reagent), 0.05 g $MgSO_4.7H_2O$ (Anedra, analytical reagent), and distilled water (Laboratory). Two different silver concentrations were selected to carry out the agar plate assays, 60 and 120 ppm. The Petri dishes were inoculated in the center with 20 mL of spore suspension of each fungus per triplicate and incubated at 28°C for 10 days. With the obtained results, the inhibition percentage (I%) was calculated according to Eq. [31]: inhibition % = $[(C - E)/C] \times 100$, where C and E correspond to the average diameter of each fungus in the control plate and on the plate with the tested solids, respectively. Three measurements of the fungal growth diameter were made in each plate, and the standard deviation was determined.

3. Results and conclusions of additives

Figure 1 shows the synthesized samples of silica with different concentrations of carbon whose images were obtained digitally. If we look at **Figure 1**, the SB sample is the one obtained with ammonium hydroxide and is taken as a control sample (it does not contain carbon), while the other images provide a light gray to dark gray coloration for higher carbon contents. For the three cases presented, the granulometry is similar when they are already dry, it is not significant compared to the SB, and only the S3B has larger granules.

The determination of the structure of the synthesized silicas was carried out by XRD. Thus the amorphous character of the synthesized materials that have only wide peaks in the 15–30° 2θ interval was confirmed and the band located around 23° 2θ was observed, which is the typical structure of this type of silica. The acid properties of the silicas measured through the potentiometric titration with *n*-butylamine were studied, which allows the evaluation of the number of acid sites and their acid strength. To interpret the results obtained, it is known that the initial electrode potential (Ei) indicates the maximum acid strength of the surface sites and the values (meq/g solid) where the plateau is reached indicate the total number of acidic sites [32]. The acid strength of surface sites can be classified according to the following ranges: very strong sites Ei > 100 mV; strong sites 0 < Ei <100 mV; weak sites −100 < Ei <0 mV, and very weak sites Ei < −100 mV, respectively [30]. It is important to clarify that this technique only indicates the trend of mass acidity of the synthesized samples. Bulk carbon has an Ei value of 37.1 mV, while silica without carbon has an Ei of 157.9 mV. It is interesting to note that the potentiometric curves have a similar shape to each other, with continuous and relatively rapid decrease in potential, which would indicate that their acidic sites are very few, regardless of the change in the amount of carbon they contain, this could that compounds that impurity carbon tend to be basic in nature. In any case, the potentiometric curves have a strong parallel with the behavior of pure silica and not of bulk carbon.

The FT-IR spectrum of the SB silica shows characteristic bands at 3748 and 3473 cm^{-1} assigned to the interactions between the hydroxyl groups on the silica surface and the water presented in the surrounding atmosphere. These bands can be related to the presence of isolated groups (Si—OH) and OH stretch bands, caused by hydrogen-bound water molecules (HOH.. H) and surface silanol groups, hydrogen-bound to water molecular (SiO—H....H_2O). The other characteristic bands that confirm the hydrophilic character of the silica are located at 968 and 1883 cm^{-1} and are directly related to the Si—O interaction of the silanol groups. At 1640 cm^{-1}, an intense band associated with the adsorption of water on the surface of the sample is also observed due to its hydrophilic nature. Bands in the range 1200–1000 cm^{-1} and 800 cm^{-1} were also detected. These interactions can be related to antisymmetric and

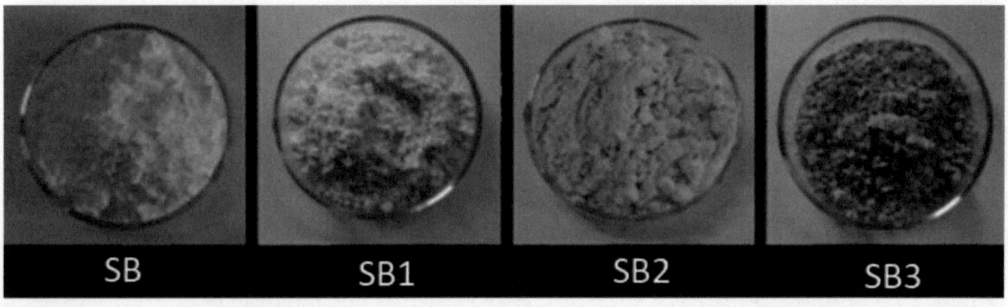

Figure 1.
Digital images of silicas.

symmetric vibration between Si–O–Si with a minimum of 1076 cm^{-1} and 801 cm^{-1}, respectively. The vibration mode that appears at 1231 cm^{-1} can be assigned to the symmetric deformation of C-H in CH_2 groups, corresponding to the residual non-hydrolyzed alkoxy groups ($–OC_2H_5$) in the silica xerogel. The characteristic interaction band was observed at 1381 cm^{-1}, which may be related to the C–H interaction of the ethyl radicals on the silica surface. These radicals can be formed as a product of condensation reactions between $Si(OH)_4$ and $Si(OC_2H_5)_4$. Carbon-containing samples show similarity to pure silica [33].

The immobilization of antimicrobial agents within multiple materials obtained by sol-gel has recently been investigated. For example, Copello et al. [34] studied the incorporation of dodecyl-di (aminoethyl) glycine in a matrix of SiO_2-xerogel for use as an antimicrobial in glasses, and Marini et al. [35] incorporated quaternary ammonium salts in an organic and inorganic hybrid coating for plastics. This methodology offers the possibility of obtaining materials of different porosity, as well as allowing the introduction of metals and other molecules through a simple impregnation, dissolution, or suspension of the metal precursors in the gel [36, 37]. In particular, several investigations are found in the literature on the use of immobilized Ag in materials obtained by sol-gel [38, 39]. Generally, materials impregnated with Ag consist of Ag ions integrated in inert ceramic, zeolite, or vitreous matrices. The sol-gel method became an effective procedure for linking organic and inorganic molecules in the same matrix and offers a unique opportunity to incorporate metal components into an organically modified inorganic matrix. The methods are entrapment, electrostatic interaction, adsorption, and covalent bonding.

The samples with Ag included are SBAg, without C, and 4% Ag that possess an Ei of 113.5 mV and S3BAg, with 10% C, and 4% Ag that showed an Ei of 67.7 mV. This could be due to the electrons of the ammonium groups that would be induced to OH more acids and may result in a redox reaction of Ag^{1+} to Ag^0. The potentiometric curves of SBAg and S3BAg are similar to the previous samples without Ag. The area under the curves is more open, indicating a greater amount of acid sites. The adsorption/desorption isotherms of N_2 corresponding to samples obtained using ammonium hydroxide as a catalyst could be included in Langmuir type II, characteristics of low porous solids, with meso- and macroporosity. Point B is where the coverage of the monolayer is complete and multilayer adsorption is about to begin. This kind of isotherm is a characteristic of nonporous solids or macroporous adsorbents. For the SBAg and S3BAg samples, the isotherms are similar which would indicate that the dopants (Ag and C) do not influence the basic hydrolysis that prevails in the synthesis of these samples. Regarding the FT-IR spectra, the samples show a shift, with respect to the SBAg. The bands are at 1182, 1094, 860, 674, and 464 cm^{-1}, but they are not substantial so that it can induce the variation of links in the siliceous network.

In the case of using ammonium hydroxide, in SEM (**Figure 2**), it can be seen that the particles of laminar morphology of the silica with acid hydrolysis become rounded. This generates a sharp decrease in the specific area and is independent of the dopants included, both in the SBAg and in the S3BAg, respectively.

It should not be forgotten when discussing this point that the sol is defined as a stable suspension of colloidal solid particles in a liquid [40]. For the existence of the sun, the colloidal particles that form it, denser than the surrounding liquid, must be small enough not to precipitate, being suspended by the repulsion of weak forces, such as those of van der Waals, or by surface charges that keep them in suspension. To meet these requirements, the particles must have sizes between 1 and 100 nm, which corresponds to the existence of 103–109 atoms per particle [41]. In the case of TEM (**Figure 3**), the rounded forms of silica and the superficial presence of Ag particles in both samples can be distinguished.

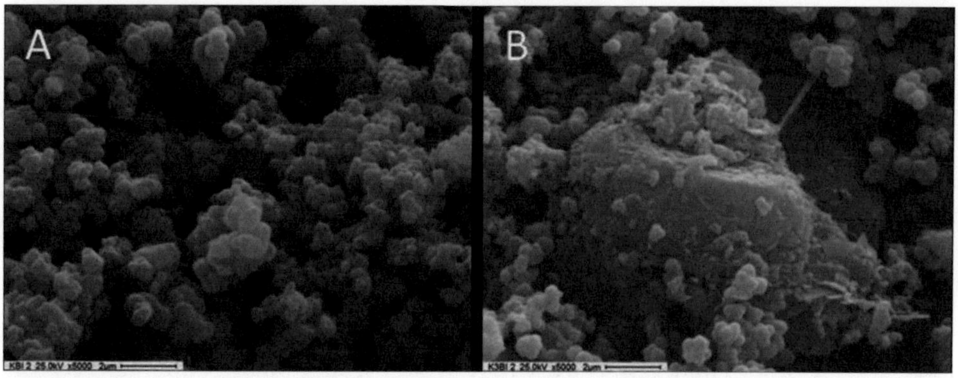

Figure 2.
SEM micrographics of samples: (A) SBAg and (B) S3Ag (5000×).

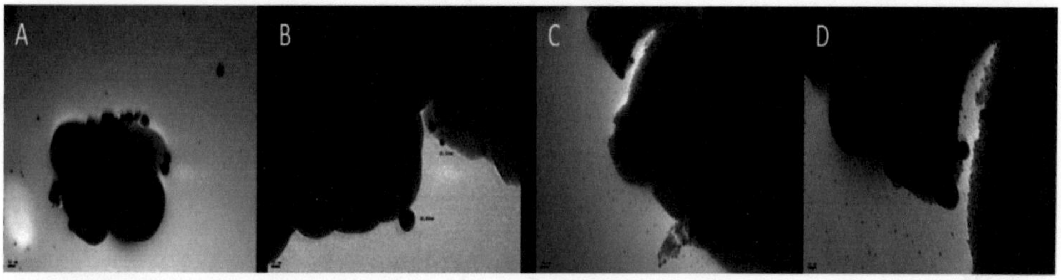

Figure 3.
TEM micrographics of samples: (A) SBAg (100,000×), (B) SBAg (270,000×), (C) S3Ag (100,000×), and (D) S3Ag (270,000×).

Babapour et al. [42] studied the inclusion of silver in a siliceous matrix through the sol-gel method and analyzed the materials by X-ray photoelectronic spectroscopy, to elucidate the chemical state of the silver nanoparticles on the surface. They observed that at 100°C, the silver particles have a high tendency to accumulate on the surface, but, at higher temperatures, they diffuse from the surface to the matrix. Also, they found that in dry samples (in air at 100°C) more than 90% of the concentration of Ag on the surface is in the Ag^0 (metallic) state. However, after treating the materials thermally at 200°C, the silver particles oxidize, presenting an increase in the surface concentration of Ag^+ and Ag^{2+}, which continues to grow up to 400°C, the results being independent of the concentration of silver in the siliceous matrix.

4. Antimicrobial fabrics

4.1 Insulation of fungi from biodeteriorated fabrics

To obtain the biodeteriorated fabric, source of the strains used in this work, samples of 100% cotton (plain weave fabric), 5 cm × 5 cm in size, previously moistened with distilled water, were exposed to accelerate the process of biodeterioration. They remained for 30 days in an indoor environment, under conditions of high relative humidity. It should be noted that this type of fabric is used in the hospital field as stretchers and oxygen tube covers, sheets, both (shirt and pants), etc. At the end of the exposure time of the samples, they were surficially decontaminated to orient the isolation to the fungal species that were growing in the fabric. According to the observations made, the isolates that presented the highest cellulolytic activity (halo ≥0.4 cm) were found to be used as bioindicators: *Aspergillus* and *Cladosporium*, respectively. On the other hand, a strain of *Chaetomium globosum* (KU936228) was also selected as a bioindicator considering that it is widely known for its cellulolytic activity (**Figure 4**).

4.2 Fabric preparation: pad-dry-cure method

Pad-dry-cure or exhaust-dry-cry is a finishing process applied to textiles to impart different finish treatments, such as waterproofing, softening, antibacterial or anti-odor finishes. The textile is passed through a water-based solution bath containing the Ag-silica additives; in this case, this method [13, 43] consisted in the inclusion of cotton fabric (4 cm × 4 cm), and the total immersion was carried out at 20°C for 10 min. Then, it was dried at 40°C for 2 h and, finally, cured for 1 h at 140°C. These impregnated fabrics were exposed against the *Chaetomium globosum* and *Aspergillus* sp. strains, to measure their antifungal activity.

4.2.1 Wash cycles

To evaluate the durability of the adhesion of the additives to the tissue, durability tests were performed against washing. Each sample was subjected to 1, 5, and 20 wash cycles of 15 min each. Each cycle consisted of placing the impregnated fabrics in a 400-ml beaker in contact with a solution of sodium lauryl sulfate 2 g/l for 15 min. Then, the fabrics were rinsed, removed with tweezers, and placed in another beaker with distilled water; this procedure was performed twice, and, finally, each cloth was rinsed again with a water slug dragging all traces of soap (**Figure 5**). The new nomenclature [44] is SBAg (KBI) and S3BAg (K3BI).

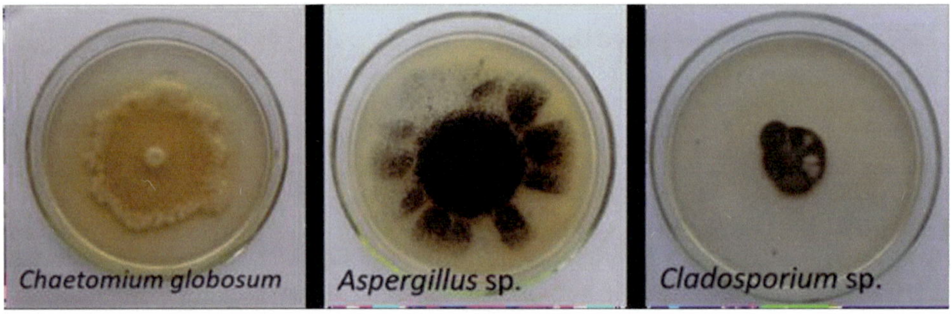

Figure 4.
Photographs of the strains used as bioindicators.

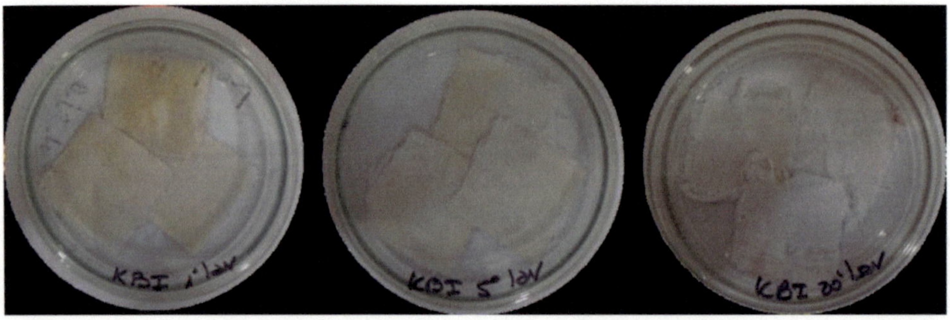

Figure 5.
Photographs of the fabric with KBI after a first wash cycle (left), 5 wash cycles (center), and 20 wash cycles (right).

The nomenclature is changed in this section that is subsequent to all changes suffered by the selected fabric. This makes the identification of pure silica-based additives based on those incorporated into the fabric simpler as explained previously.

4.3 Evaluation of the antifungal activity of fabrics

The antifungal activity of the fabrics treated with the modified silicas was estimated with the bioindicators: *Aspergillus* sp. and *C. globosum* (KU936228) according to the standard modified method DIN 53931390. The culture medium used consists of 1 g of KH_2PO_4, 1 g of KNO_3, 0.5 g of $MgSO_4.7H_2O$, 0.5 g of KCl, 0.2 g of glucose, 0.2 g of sucrose, and 15 g of agar per 1 l of distilled H_2O. It is a less nutritious culture medium, allowing more delicate colony growth and an easier evaluation of the antifungal activity of the fabric. About 100 μl of the spore suspension (inoculum) previously obtained was inoculated, spread with the Drigalski spatula to obtain a homogeneous lawn of the strain, and incubated in an oven at 28°C for 24 h. Subsequently, the impregnated fabrics (4 cm × 4 cm) were sterilized by UV radiation and placed in the center of the previously grown plate working in the laminar flow.

Then, they were incubated in an oven 28°C for 14 days. After that time, the antifungal activity was determined in terms of mycelial growth on the surface of the cotton fibers and the intensity of the sporulation. To ensure statistical validity, the test was performed in triplicate.

4.3.1 Results

Analyzing the data after 20 wash cycles, some of the antifungal activities is lost. Both SBAg and S3BAg samples have no noticeable differences in growth inhibition, achieving only a dispersed growth of between 5 and 10% (eye

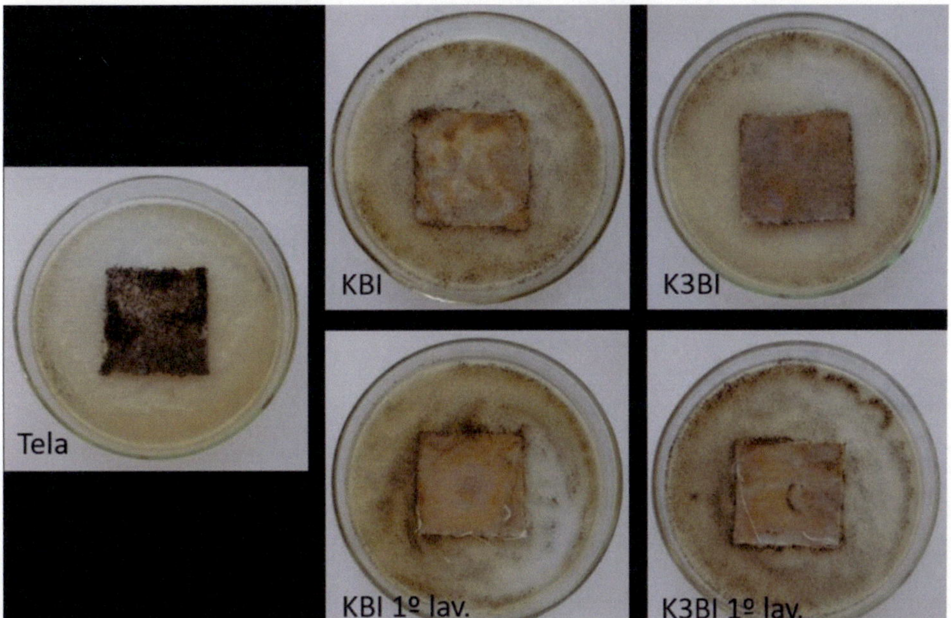

Figure 6.
Antifungal test of the fabrics impregnated with the pad-dry-cure method against C. globosum.

observation). As can be seen in **Figure 6**, which shows the photographs of the test against *Chaetomium globosum*, there is a greater sporulation concentrated in the control fabric, while in the other fabrics containing Ag there is only to a lesser extent on the edges of the fabric. The nomenclature is as follows [44]: control (Tela), SBAg 1 wash cycle (KBI 1 lav.), and S3BAg 1 wash cycle (K3BI 1 lav.).

Figure 7 shows SEM micrographs of the fabrics tested. Here the difference in growth in the control fabric with respect to the fabrics with KBI and K3BI versus *C. globosum* is noticeable, although the KBI gives less growth than the fabric containing K3BI. It can be seen that KBI has lower growth than K3BI, which has poor and scattered growth. The micrographs, which are shown by way of example, can be seen

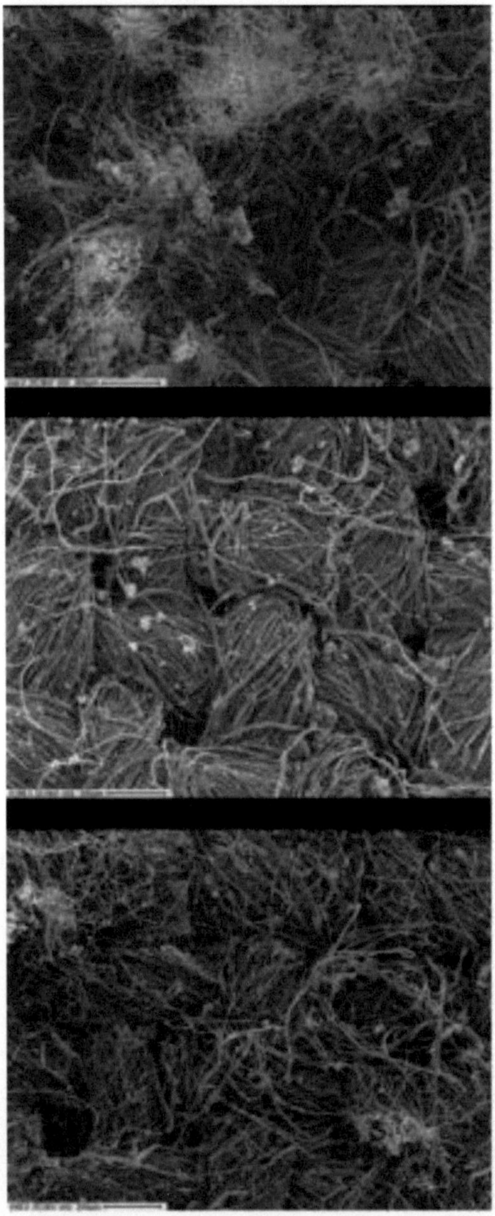

Figure 7.
SEM micrographs of the control fabric (above) and the fabrics impregnated with the pad-dry-cure method KBI and K3BI (medium and down), tested against C. globosum.

again that *Aspergillus* sp. (**Figures 8** and **9**) has scattered and weakened specialized hyphae (conidiophores), in fabrics containing KBI and K3BI, with respect to the control fabric where they are more abundant and with normal characteristics.

As a conclusion of this section regarding the additives impregnated with the fabrics, these samples were synthesized with basic hydrolysis, the KBI does not contain carbon, and there is only impregnation of Ag in the sample of the previous stage. For the K3BI, there is presence of C together with Ag. If the activity is compared, these samples gave good results. Regarding the washing cycles, there is no difference between the samples for the fungi tested, there may be loss of the additive with the number of washes, but there is no variation between 1 cycle and 20 cycles, which leads to good adhesion of the additives to the fabric, that is, the method tested has a good rating to continue using it.

4.4 Evaluation of the antibacterial activity of fabrics

For the test of antimicrobial activity, a first general classification of the method to be used is carried out depending on the type of evaluation of the population of microorganisms. Reduction in intimate contact with an agar culture medium inoculated with the test bacteria (DIN EN ISO 20645-2001, AATCC 147). If diffuse or leaching antibacterial activity is present, it will be possible to observe a clear area around the treated sample compared to the surrounding bacterial growth zone and the untreated control sample after the same contact time. However, this method cannot be applied to nondiffusible or fixed antimicrobial substances [45].

4.4.1 Agar diffusion method (SN 195920-1992)

To study the antibacterial efficacy of the impregnated fabrics, the agar-based diffusion method was performed (SN 195920-1992). Bacterial strains for test in *E. coli* and *S. aureus* (**Figure 10**) were selected to be abundant in the

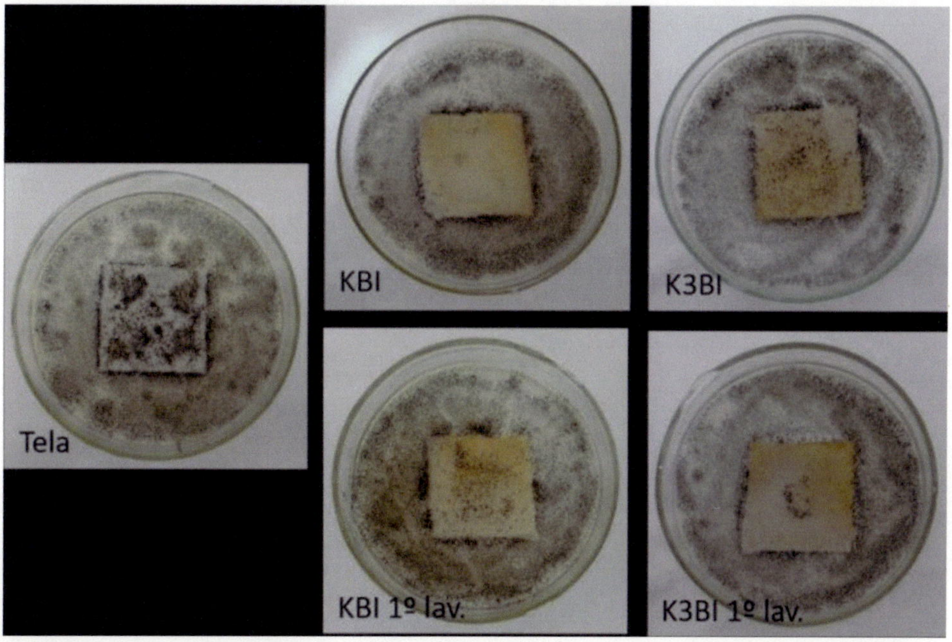

Figure 8.
Antifungal test of fabrics impregnated with the pad-dry-cure method against Aspergillus *sp.*

environment and be related to pathologies affecting human health. The culture medium used is BVAC: 5 g NaCl, 5 g yeast extract, 10 g casein peptone, and 15 g of bacteriological agar 1 l of distilled water. Then, plates were prepared with 15 ml of the culture medium BVAC and inoculated with the inoculum previously prepared, which spread throughout the plate with sterile swabs. Finally, the fabrics were added treated and untreated. The plates were incubated for 24 h at 37°C [14, 46].

Figure 9.
SEM micrographs of the control fabric (above) and the fabrics impregnated with the pad-dry-cure method KBI and K3BI (medium and down) and the fabrics impregnated with the pad-dry-cure method, tested against Aspergillus *sp.*

The inoculum was made from 24-h cultures that were in an oven at 37°C. Suspensions with physiological solution were obtained by adjusting the turbidity to 0.5 of the McFarland scale (1.5 × 108 Ufc/ml). A dilution was then made to obtain a bacterial suspension of 1.5 × 106. After the incubation period of the plates inoculated with the selected strains, the zone of inhibition (ZOI) was recorded. The results were obtained from the average of four measurements taken for each triplicate. In addition, the standard deviation between measurements was determined. **Figures 11** and **12** show the photographic records of the trial and those observed through the magnifying glass of the fabrics against *E. coli*. It can be seen that there is an inhibition halo that is identified as a space adjacent to the fabric (transparent culture medium). There are no noticeable differences in the measures of the ZOI of the fabrics that contained the additives in spite of the washing cycles; therefore it can be concluded that the impregnation method has high durability against washing. Taking into account the values of ZOI, 0.6 ± 0.2 for KBI and 0.7 ± 0.2 for K3BI, it can be said that only at 20 wash cycles there is a decrease in the antibacterial effect.

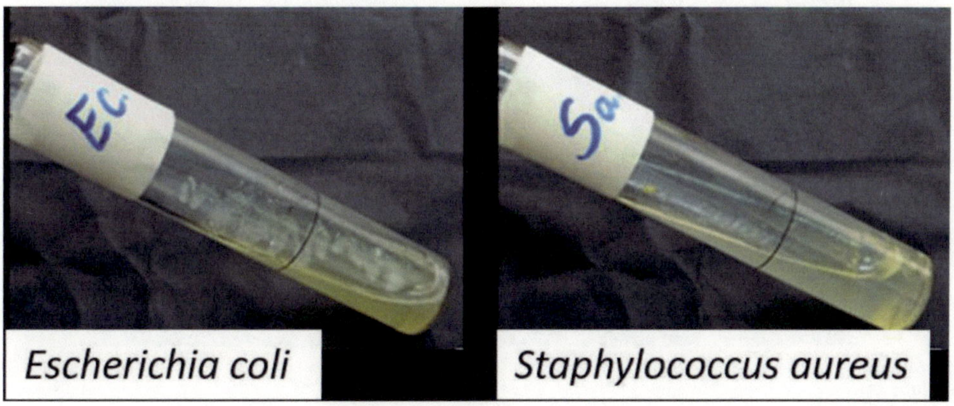

Figure 10.
Photographs of bacterial strains used in the assay.

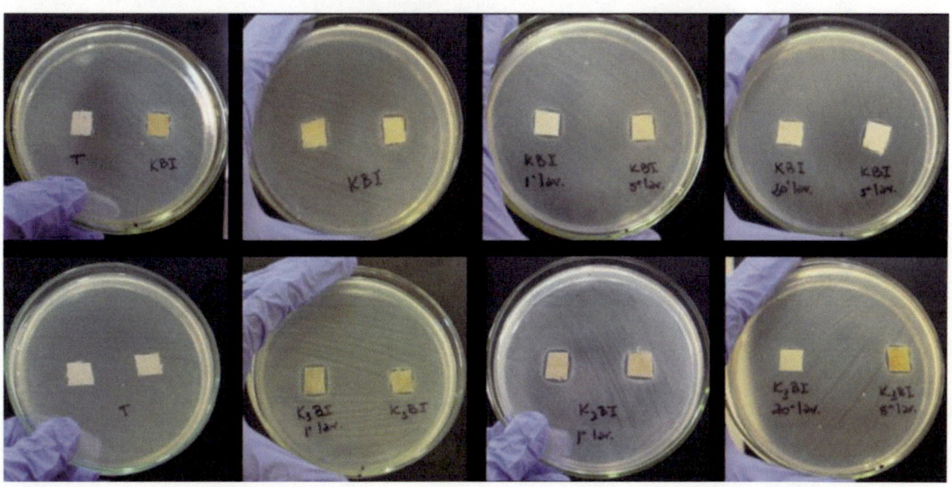

Figure 11.
Antibacterial test of fabrics impregnated with the pad-dry-cure method against Escherichia coli.

The photographs of the agar diffusion test against *S. aureus* are shown in **Figures 13** and **14**, in which a halo of inhibition is observed for the fabrics impregnated with the biocide while in the control fabric there is growth throughout the plate.

The photographs obtained through the magnifying glass clearly show the interface of fabric-culture medium-bacterial growth for fabrics with biocide, thus affirming their inhibitory effect. With respect to the washing cycles, they have the same tendency as described for *E. coli*, producing a slight decrease in activity only in the washing cycle number 20.

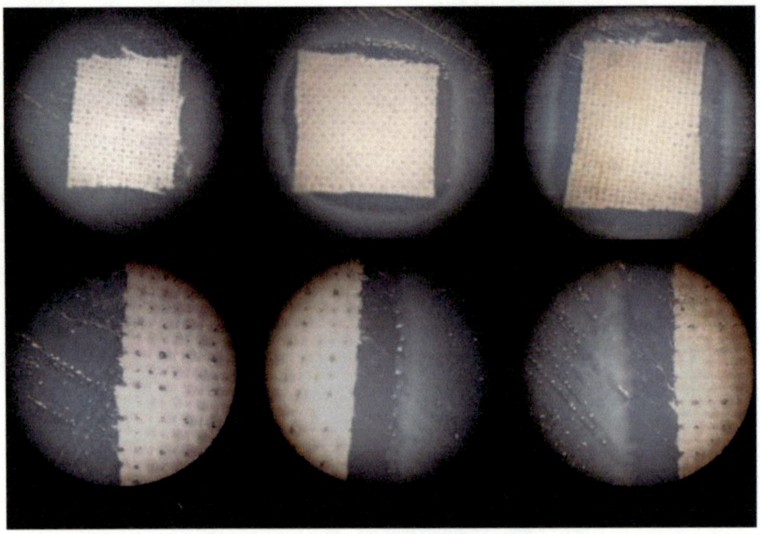

Figure 12.
Images observed with magnifying glass of the control fabric and the fabrics impregnated with the pad-dry-cure method, tested against Escherichia coli.

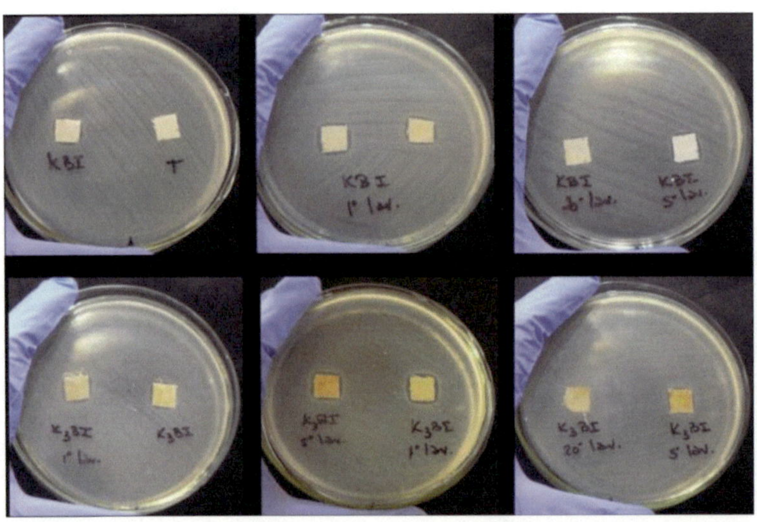

Figure 13.
Antibacterial assay of fabrics impregnated with the pad-dry-cure method against Staphylococcus aureus.

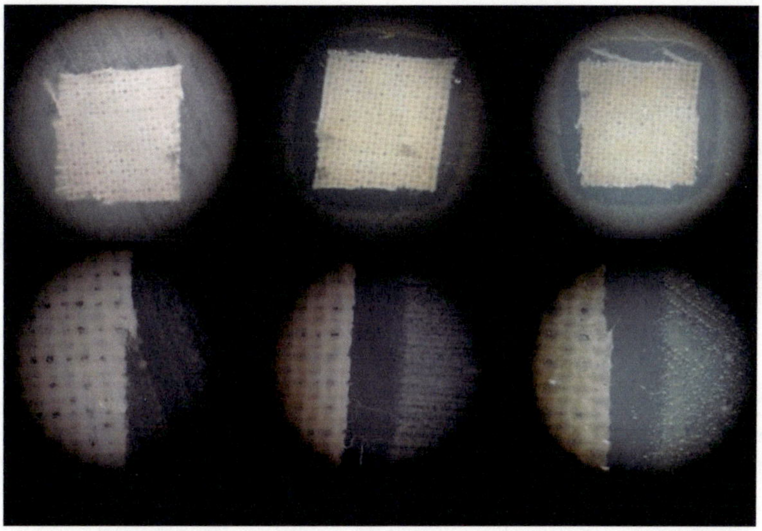

Figure 14.
Images observed with magnifying glass of the control fabric and the fabrics impregnated with the pad-dry-cure method, tested against Staphylococcus aureus.

5. Conclusions

As a closing of this chapter, it can be concluded that antimicrobial fabrics were obtained from the pad-dry-cure method, using the samples synthesized on silica base with C and Ag. The cotton cloth used was evaluated in antifungal tests with strains *C. globosum* and *Aspergillus* sp., according to the modified standard method DIN 5393, and by antibacterial assays through the agar-based diffusion method (SN 195920-1992), against *E. coli* and *S. aureus*.

On the other hand, the results were good for both the inhibition of fungal and bacterial strains. In addition, there was a high degree of persistence of the additives after the wash cycles before antimicrobial tests, with inhibition being recorded up to 20 cycles.

Since most of the work focuses on the inhibitory effect of Ag against bacteria, the mechanisms of inhibition or lethality of surfaces with antifungal compounds are poorly understood. In recent years there has been an effort to increase research on antimicrobial fabrics, but the vast majority of these publications focus on bacterial research and only some conduct resistance tests against fungi. If the number of investigations in general is reviewed, a marked increase is observed in the last 5 years. However, the number of investigations on antibacterial fabrics is higher than the number of investigations on antifungal fabrics. However, taking into account the importance of fungal infections in human health, considering that the incidence of fungal infections is increasing and the associated health costs are high, studies should also focus on this field and interest on antifungal fabrics should be growing.

The antimicrobial properties of silver have been known for many years, but recently it has begun to understand the mechanisms by which silver inhibits the growth of microorganisms, although much remains to be investigated. There are several investigations about the biocidal effect and mode of action of silver ions against bacteria, especially against *E. coli*; however, the antifungal effects and the mode of action of nano-Ag against fungi have not yet been studied in depth, and it remains unknown.

Some investigations assume that Ag atoms bind to thiol groups (–SH) in enzymes and subsequently cause them to be deactivated by denaturation. Stable S–Ag bonds would be generated with compounds that contain thiol in the cell membrane and that would be involved in the generation of transmembrane energy and in the transport of ions [43]. The result would be a loss of fluids and electrolytes from microorganisms, which are dried and shrink. In literature it was shown that the cells suffered great damage when contacted with Ag, a treatment that finally destroyed the cell wall and the cell membrane. Damage to the cell membrane could lead to cell cytoplasm filtration, which would result in dehydrated and shrunken cells [43].

The method to obtain the carbon used in these samples comes from a process that can be applied to the treatment of residues with high metal contents, which resemble a non-sulfide mineral, for example, batteries, computer waste, slags from the metallurgical industry, etc. The acid medium can be generated in situ in biopiles directly on the mineral or residue, or produced in bioreactors where the microorganisms are grown, and then put in contact with the mineral or residue; the latter is the case studied in this work. In recent years, scientific papers have been reported on this process applied to the recovery of metals from different wastes, among others, batteries and batteries, bibliographic references previously placed, although no existing commercial facilities have been reported to date [47, 48].

It can be concluded, in general terms, that the proposed objectives have been achieved, since antimicrobial additives were synthesized through a simple and rapid method of obtaining, such as the sol-gel method, which allowed the inclusion of the biocide, Ag, in oxidic matrices. They could be incorporated effectively in the preparation of antimicrobial fabrics.

Acknowledgements

This work was supported by the National Scientific and Technical Research Council (CONICET) and the National University of La Plata (UNLP).

Conflict of interest

The authors declare no conflict of interest.

Author details

Katerine Igal and Patricia Vázquez*
Center for Research and Development in Applied Sciences (CINDECA), National Scientific and Technical Research Council (CONICET), La Plata University, La Plata, Argentina

*Address all correspondence to: vazquez@quimica.unlp.edu.ar

References

[1] Balter M. Clothes make the (Hu) Man. Science. 2009;**325**(5946):1329. DOI: 10.1126/science.325_1329a

[2] Byko M. From electric corsets to self-cleaning pants: The materials science and engineering of textiles. Journal of The Minerals, Metals & Materials Society. 2005;**57**(7):14-18. JOM is published monthly in partnership with Springer (ISSN: 1047-4838)

[3] Gao Y, Cranston R. Recent advances in antimicrobial treatments of textiles. Textile Research Journal. 2008;**78**:60-72. DOI: 10.1177/0040517507082332

[4] Kim YH, Sun G. Dye molecules as bridges for functional modifications of nylon: Antimicrobial functions. Textile Research Journal. 2000;**70**:728-733. DOI: 10.1177/004051750007000811

[5] Available from: Fed/BIO/Products/Brand/reputex. 2007. www.archchemicals.com/

[6] Payne S. Antimicrobial superfinish and method of making. 2004. United States Patent Application US20040077747A1

[7] Lim SH, Hudson SM. Review of chitosan and its derivatives as antimicrobial agents and their uses as textile chemicals. Journal of Macromolecular Science: Part C: Polymer Reviews. 2003;**43**:223-269. DOI: 10.1081/MC-120020161

[8] Sun G, Xu X. Durable and regenerable microbiocidal textiles. 1999. United States Patent US5882357A

[9] Huang LK, Sun G. Durable and regenerable antimicrobial cellulose with oxygen bleach: Concept proofing. AATCC Review. 2003;**3**:17-21

[10] El-Rafie MH, Ahmed HB, Zahran MK. Characterization of nanosilver coated cotton fabrics and evaluation of its antibacterial efficacy. Carbohydrate Polymers. 2014;**107**:174-181. DOI: 10.1016/j.carbpol.2014.02.024

[11] Mahltig B, Fiedler D, Böttcher H. Antimicrobial sol–gel coatings. Journal of Sol-Gel Science and Technology. 2004;**32**:219-222. DOI: 10.1007/s10971-004-5791-7

[12] Mahltig B, Fiedler D, Simon P. Silver-containing sol–gel coatings on textiles: Antimicrobial effect as a function of curing treatment B. The Journal of The Textile Institute. 2011;**102**(9):739-745. DOI: 10.1080/00405000.2010.515730

[13] Tomšič B, Simončič B, Orel B, Zerjav M, Schroers H, Simončič A, et al. Antimicrobial activity of AgCl embedded in a silica matrix on cotton fabric. Carbohydrate Polymers. 2009;**75**:618-626. DOI: 10.1016/j.carbpol.2008.09.013

[14] Budama L, Acar Çakır B, Topel Ö, Hoda N. A new strategy for producing antibacterial textile surfaces using silver nanoparticles. Chemical Engineering Journal. 2013;**228**:489-495. DOI: 10.1016/j.cej.2013.05.018

[15] Robles-García MA, Rodríguez FF, Márquez-Ríos E, Barrera-Rodríguez A, Aguilar-Martínez J, Del-Toro-Sánchez CL. Aplicaciones biomédicas, textiles y alimentarias de nanoestructuras elaboradas por electrohilado. Revista de Ciencias Biológicas y de la Salud. 2014;**16**(2):44-52. DOI: 10.18633/bt.v16i2.46

[16] Dao H, Lakhani P, Police A, Kallakunta V, Ajjarapu SS, Wu K-W, et al. Mini-review: Stability of pharmaceutical excipients. In: Narasimha Murthy S, Repka MA, editors. Microbial Stability of Pharmaceutical and Cosmetic

Products. AAPS PharmSciTech; 2017. DOI: 10.1208/s12249-017-08751

[17] Palza H, Escobar B, Bejarano J, Bravo D, Diaz-Dozque M, Perez J. Designing antimicrobial bioactive glass materials with embedded metal ions synthesized by the sol-gel method. Materials Science and Engineering. 2013;**33**:3795-3801. DOI: 10.1016/j.msec.2013.05.012

[18] Sørensen G, Nielsen AL, Pedersen MM, Nygaard SD. Controlled release of biocide from silica microparticles in wood paint. Progress in Organic Coating. 2010;**68**:299-306. DOI: 10.1016/j.porgcoat.2010.03.009

[19] Arreche A, Igal K, Bellotti N, Deyá C, Vázquez P. Functionalized zirconia compounds as antifungal additives for hygienic waterborne coatings. Progress in Organic Coatings. 2019;**128**:1-10. DOI: 10.1016/j.porgcoat.2018.12.004

[20] Bosecker K. Bioleaching: Metal solubilization by microorganisms. FEMS Microbiology Reviews. 1997;**20**:591-604. DOI: 10.1111/j.1574-6976.1997.tb00340.x

[21] Ehrlich HL. Geomicrobiology. Cuarta edición ed. New York: Dekker/CRC Press; 2002

[22] Olson GJ, Brierley JA, Brierley CL. Progress in bioleaching: Applications of microbial processes by the minerals industries. Applied Microbiology and Biotechnology. 2003;**63**:249-257. DOI: 10.1007/s00253-003-1404-6

[23] Rawlings DE. Heavy metal mining using microbes. Annual Review of Microbiology. 2002;**56**:65-91. DOI: 10.1146/annurev.micro.56.012302.161052

[24] Rohwerder T, Jozsa PG, Gehrke T, Sand W. Bioleaching. In: Bitton G, editor. Encyclopedia of Environmental Microbiology. Vol. 2. New York: Wiley; 2002. pp. 632-641. DOI: 10.1007/1-4020-5589-7

[25] Bellotti N, Romagnoli R, Quintero C, Domínguez-Wong C, Ruiz F, Deyá C. Nanoparticles as antifungal additives for indoor water borne paints. Progress in Organic Coating. 2015;**86**:33-40. DOI: 10.1016/j.porgcoat.2015.03.006

[26] Samson RA, Hoekstra ES, Frisvad JC. Introduction to Food and Airborne Fungi. 7th ed. Netherlands: CBS; 2004

[27] Olaf CGA, Samson RA. Fundamentals of Mold Growth in Indoor Environments and Strategies for Healthy Living. Netherlands: Wageningen Academic Publishers; 2011. ISBN-13: 978-9086861354, ISBN-10: 9086861350

[28] Bellotti N, Salvatore L, Deyá C, Del Panno MT, del Amo B, Romagnoli R. The application of bioactive compounds from the food industry to control mold growth in indoor waterborne coatings. Colloids and Surfaces. B, Biointerfaces. 2013;**104**:140-144. DOI: 10.1016/j.colsurfb.2012.11.037

[29] Espinosa DCR, Bernardes AM, Tenório JAS. Brazilian policy on battery disposal and its practical effects on battery recycling. Journal of Power Sources. 2004;**137**:134-139

[30] Arreche R, Bellotti N, Blanco M, Vázquez P. Synthesis and characterization of zirconium oxides for use as antimicrobial additives in paints. Procedia Materials Science. 2015;**9**:627-634. DOI: 10.1016/j.mspro.2015.05.039

[31] Romanelli R, Vázquez P, Pizzio L, Quaranta N, Autino J, et al. Phenol tetrahydropyranylation catalyzed by silica-alumina supported heteropolyacids with Keggin structure.

Applied Catalysis A: General. 2004;**261**:163-170. DOI: 10.1016/j.apcata.2003.10.043

[32] Ruiz D, Romanelli G, Vázquez P, Autino J. Preyssler catalyst: An efficient catalyst for esterification of cinnamic acids with phenols and imidoalcohols. Applied Catalysis A: General. 2009;**374**(1-2):110-119. DOI: 10.1016/j.apcata.2009.11.037

[33] Salazar D, Catalina A. Desarrollo de Materiales a Base de sílice Para la adsorción de Compuestos orgánicos (Development of Silica Containing Materials for the Adsorption of Organic Compounds) [Maestría thesis]. Sede Manizales: Universidad Nacional de Colombia; 2011

[34] Copello GJ, Teves S, Degrossi J, D'Aquino M, Desimone MF, et al. Antimicrobial activity on glass materials subject to disinfectant xerogel coating. Journal of Industrial Microbiology & Biotechnology. 2006;**33**:343-348. DOI: 10.1007/s10295-005-0066-z

[35] Marini M, Bondi M, Iseppi R, Toselli M, Pilati F. Preparation and antibacterial activity of hybrid materials containing quaternary ammonium salts via sol–gel process. European Polymer Journal. 2007;**43**:3621-3628. DOI: 10.1016/j.eurpolymj.2007.06.002

[36] Strek W, Maruszewski K, Jasiorski M, Lukowiak E, Bryja L. Spectroscopic properties of CdS nanoparticles embedded in sol-gel silica glasses. Optica Applicata. 1999;**29**:401-405. DOI: 10.1016/j.eurpolymj.2007.06.002

[37] Samuneva B, Dimitriev Y, Dimitrov V, Kashchieva E, Encheva G. Silica gels and gel glasses containing silver and platinum metal particles. Journal of Sol-Gel Science and Technology. 1998;**13**:969-974. DOI: 10.1023/A:1008695812143

[38] Bellantone M, Coleman NJ, Hench L. Bacteriostatic action of a novel four-component bioactive glass. Journal of Biomedical Materials Research. 2000;**51**:484-490. DOI: 10.1002/1097-4636(20000905)51:3<484::aid-jbm24>3.0.co;2-4

[39] Jeon HJ, Yi SC, Oh SG. Preparation and antibacterial effects of Ag–SiO_2 thin films by sol-gel method. Biomaterials. 2003;**24**:4921-4928. DOI: 10.1016/s0142-9612(03)00415-0

[40] Livage J, Lemerle J. Transition metal oxide gels and colloids. Annual Review of Materials Science. 1982;**12**:103-122. DOI: 10.1146/annurev.ms.12.080182.000535

[41] Hiemenz PC. Principles of Colloid and Surface Chemistry. New York: Marcel Dekker; 1977. DOI: 10.1002/pol.1984.130220910

[42] Babapour A, Akhavan O, Azimirad R, Moshfegh AZ. Physical characteristics of heat-treated nanosilvers dispersed in sol-gel silica matrix. Nanotechnology. 2006;**17**:763-771. DOI: 10.1088/0957-4484/17/3/025

[43] Mahltig B, Fiedler D, Fischer A, et al. Antimicrobial coatings on textiles-modification of sol-gel layers with organic and inorganic biocides. Journal of Sol-Gel Science and Technology. 2010;**55**:269-277

[44] Igal K. Telas Antimicrobianas Impregnadas con Partículas de Ag o Zn Reciclado Incluidas En Matrices Silíceas Modificadas. Universidad Nacional de La Plata; 2019

[45] Testing for antimicrobial activity in textiles—Quick overview. AATCC Review. ISSN 1532-8813 (Online). ISSN:2330-5525 (Print)

[46] Balakumaran MD, Ramachandran R, Jagadeeswari S, Kalaichelvan PT.

In vitro biological properties and characterization of nanosilver coated cotton fabrics: An application for antimicrobial textile finishing. International Biodeterioration and Biodegradation. 2016;**107**:48-55 ISSN: 0964-8305

[47] Falco L. Optimización de Un Proceso biohidrometalúrgico de recuperación de Metales de Pilas Alcalinas y de Cinc-carbón Agotadas. Universidad Nacional de La Plata; 2013, oai:sedici.unlp.edu.ar:10915/29548

[48] Gallegos V. Estudio del Uso de óxidos de Metales de transición Obtenidos a Partir del Reciclado de Pilas Como Fases Activas y Soportes en la remoción de Compuestos orgánicos volátiles y Otras Aplicaciones. Universidad Nacional de La Plata; 2017. Available from: https://doi.org/10.35537/10915/65857; http://sedici.unlp.edu.ar/handle/10915/65857

Chapter 2

The Waste Problem of Antimicrobial Finishing

Candan Akca

Abstract

Growing awareness of health and hygiene has increased the demand for bioactive or antimicrobial textiles. As a result the global market of antimicrobial textile products has been growing day by day. Antimicrobial finishing protects the wearer from microorganisms for aesthetic, hygiene, or medical reasons and protects the textile from biodeterioration caused by mold, mildew, and fungi. Antimicrobial textile products have crucial functions such as ensuring hygienic conditions and preventing spread of diseases especially crowded places like hospitals, baby nurseries, and barracks. However the antimicrobial agents used in antimicrobial finishing have adverse effects (toxic, allergic, and carcinogenic) on wearer and all the living organisms during the antimicrobial textile product's production and serving life. The effects of finish content released in waste water bath and the release of finish content to surrounding skin of user and its contact/inhalation/consumption by other living species require studies. The contamination substances in the sludge may be consumed by the cultivated plants, eventually becoming component in feed and food. Contamination is also possible in water and soil organisms. The chapter is about the waste problem of the antimicrobial finishing (especially metal-based antimicrobial finishing and triclosan-based antimicrobial finishing).

Keywords: antimicrobial finishing, heavy metal, silver, triclosan, waste, environment, pollution, toxicology

1. Introduction

The foundations of the textile industry were laid in Britain; spinning and weaving technologies developed here. However, in the aftermath of this development, in the nineteenth century, textile production shifted to Europe and North America following pace with the industrial modernization in these regions. In the preceding years, almost all countries have realized industrialization and development processes via the textile industry. Countries undergoing the development process continue to produce more traditional textile products, whereas countries that have already completed their development processes and have achieved an advanced level of technology continue to produce high-tech technical textiles [1].

While the textile industry was initially in a traditional position that met basic requirements, such as yarn and fabric production, clothing, and home textiles, the development of technology over time and the increase in human needs have resulted in the industry being much more technological and functional today as a result of diversification [1].

Looking at the global textile market volume in 2015 and beyond, it is observed that it reached $667.5 billion in 2015. Europe accounted for 54.6% of this volume and the Asia-Pacific region 20.6%. When the 2018 data is examined, the global textile market volume is estimated to be $858 billion, up to 5% from 2019, and estimated to reach $1.207 billion by 2025 [1, 2]. It is also estimated that interest in high-tech textile products will continue to increase in the coming years and that the market in this field will continue to grow in Europe (especially Germany, France, and Italy) [2].

In recent years, consumers' desire to feel comfort, be hygienic, feel good and control odor, and be protected from microorganisms has led to the rapid growth of the market of antimicrobial textiles [1, 3, 4]. The current uses of antimicrobial textiles range from outdoor applications such as tents, tarpaulins, awnings, blinds, parasols, sails, and waterproof clothing to indoor applications such as shower curtains and mattress ticking. They are also used in some consumer textiles such as sportswear, T-shirts, and socks and also in medical purpose such as masks, surgical clothing, wound dresses, and bandages [5]. Global antimicrobial textile market volume in 2019 reached $9468 million. And it is estimated to reach $12,313 million in 2024 [6]. According to the 2015 data, the market volume of global finishing chemicals is 1.14 million tons, and there is an increase of 6.1% each year by 2025. A significant portion of this amount consists of antimicrobial finishing agents [7]. The volume of the global wet wipe and wet napkin market is thought to have the potential to increase by $5.75 billion between 2020 and 2024 [8]. And by 2016, the volume of diapers will be $46.50 billion and is estimated to reach $67.46 billion by the end of 2022 [9].

By 2025, the world's population is estimated to be 8.2 billion. Growing world populations, rising living standards, and fast fashion trends are causing the global textile industry to grow day-by-day. This also means large amounts of raw materials and resource usage, ultimately producing pollution and a high rate of waste [1, 10].

In this section, the issue of antimicrobial textile production (especially metal-based antimicrobial textile production, triclosan-based antimicrobial textile production) and subsequent product life spans are investigated.

2. Textile and waste

The word waste is defined as things that people do not need and want to get rid of. Waste according to their physical form can be classified as solid, liquid, and gas. It is also possible to classify waste according to its original uses (packaging waste, textile waste, etc.), according to its materials (glass, metal, fiber, etc.), according to its physical characteristics (recyclable, composite, fuel, fertilization, etc.), according to its origin (domestic, commercial, industrial, agricultural, etc.), and according to its safety level (dangerous or hazardous) [9]. Household waste and commercial waste are classified together as solid municipal wastes. Excessive and unnecessary consumption in all areas of daily life increases the burden of a clogged-up world [10].

European Union member states have targeted a 50% reduction of municipal waste by 2020 through reuse and recycling [10].

In the textile industry, during the production of textile products, high amounts of solid, liquid, and gas in the form of wastes are produced as well as during the lifetime of textile products by the consumer and after the end of its lifespan [1, 10, 11].

Aged textile processes, which include pre-finishing, dyeing, printing, and finishing processes, usually consist of chemical applications, fixation, washing, and drying steps. In particular dyeing and finishing processes are processes where the highest amount of water is used [1, 11]. In textile production, a very large amount of water, chemicals, dyeing, and auxiliary chemicals are used. Therefore, textile wastewater is contaminated with these substances, has alkaline at high concentrations, is sharply scented, has the need for high biological oxygen (BOD) and for chemical oxygen (COD), and contains highly dissolved solids if it is not properly removed, which can cause environmental complications [1, 11].

Textile products other than disposable products are repeatedly exposed to washing, drying, ironing, and dry cleaning during their lifetime [12]. Wastewater contaminated with detergent, stain remover, and softener in washing baths are toxic to marine creatures [13]. With each washing, the active finishing chemicals applied on the textile product also leave the textile surface and pass on to the washing water subsequently increasing the waste load. Active finishing chemicals can leave the textile surface not only with bathing but also when faced with bodily fluids during use. This condition can cause itching, skin sensitivity, and allergies in people with sensitive skin [14].

The average life expectancy of textile products is 2 years, and then they continue to be waste loads by being stored in landfills. The amount of textile waste that has completed its life span is 10.5 million tons per year in the United States, 350,000 tons in the UK, and 287,000 tons in Turkey [10]. In particular, some studies and trends have been initiated to evaluate textile products that have completed their life spans in the United States and Europe. These studies can be summarized as recycling, reuse, energy production, second-hand clothing trends, vintage clothing trends, and slow fashion trends [10, 15]. According to 2009 data, only 15% of the textile products that have completed their life in the United States are utilized through recycling or donation, and the remaining 85% are left to solid waste landfills. However, it is thought that it is possible to utilize up to 95% with successful waste management [10].

The textile industry also produces waste in gas form, causing air pollution. Especially in spinning and weaving processes, a large amount of dust and sublimates are emitted into the operating environment. This condition can cause respiratory diseases and chronic lung diseases in workers [10].

From an environmental point of view of the textile industry, energy consumption, gas emissions, solid waste, and odor problems are also important issues, but the main problem is the chemical waste load produced in large quantities of wastewater and the chemicals in the wastewater [10].

2.1 Toxic or hazardous waste in the textile industry

Toxic or hazardous waste is waste that is dangerous for the environment and human health or has the potential to create harmful effects. Toxic and hazardous waste can occur in the form of solid, liquid, gas, and sludge as a result of various industrial production activities [16].

There are many toxic and dangerous chemicals in textile wastewater caused by different production processes [10]. Some of these are as follows [10]:

- Chlorinated solvents: chlorinated solvents are used in many processes such as bleaching, scouring, and dyeing in the textile industry. They are known to have allergic, carcinogenic, and toxic properties for human and environmental health.

- Hydrocarbon solvents-aliphatic hydrocarbons: hydrocarbons of organically structured compounds consisting of aliphatic compounds and carbon and hydrogen elements are aliphatic. They can be straight-chained, branched, or ringed and are divided into two: saturated and unsaturated. They are flammable and have sultry properties. They are known to cause nervous system diseases and cancer.

- Hydrocarbon solvents-aromatic hydrocarbons: it is very difficult to purify textile wastewater from aromatic hydrocarbons. Aromatic hydrocarbons are not easily dissolved in water. Most aromatic hydrocarbons stick to solid particles, settling in lake and riverbeds and blending into groundwater. These compounds are known to cause cancer in the long term.

- Oxygenated solvents (alcohols/glycollics/ethers/esters/ketones/aldehydes): oxygenated solvents are solvents with a high solvent feature containing an oxygen molecule. These solvents (methanol, ethanol, propane, ethylene glycol, etc.) are widely used in textile processes. They are harmful to both human health and all flora and fauna. Exposure to high amounts of these compounds can lead to sudden deaths. Prolonged exposure can cause blindness, irregular heartbeat, and damage to the kidney and lungs. Some of these compounds are in the carcinogenic category for humans. Glycol ethers can cause developmental impairment in the fetus and infertility in men. Regular exposure to these solvents can cause memory and hearing loss, depression, headache, coordination disorders, and skin disorders. Exposure to the vapors of these solvents can cause ailments such as asthma or shortness of breath.

- Grease and oil contaminated waste: grease can be animal-based, oil-based, and synthetic-based. Wastewater contaminated with grease is toxic to marine life in the long run.

- Used oils: some of the oils used in textile processes are carcinogenic to human health if they are in physical contact with humans or digested.

- Dye materials and pigments containing harmful substances: the presence of dye substances and treatment of textile wastewater are serious problems because most dye materials are stable and are not easy to parse with traditional treatment methods. The chemical structures and contents of the dyes have an effect on toxicity sites:

- Organohalogens: pigments can contain fluorocarbon, chlorocarbon, bromocarbon, or iodo-carbon bond and contains toxic elements such as lead, cadmium, mercury, valve, chromium, cobalt, nickel, arsenic, etymon, and selenium and are toxic and dangerous.

- Organic compounds (such as benzyte, methane, paraffin) are made up of carbon and hydrogen elements; they are found in coal, crude oil, natural gas, and vegetables. Hydrocarbons, pesticides, dyes, and plastics are the cornerstone of numerous product groups.

3. Textile antimicrobial treatments

Antimicrobial finishing applied to textile material should be effective against microorganisms as well as meet a number of requirements including the fact that antimicrobial finishing is suitable for the textile process; is resistant to washing, dry cleaning, and hot press; and is not harmful to the environment [17].

Different antimicrobial methods of finishing may be preferred depending on the genus, structure, surface characteristics, and usage area of textile material. Antimicrobial finishing can be carried out during the phase of finishing procedures, as well as the application of antimicrobial agents into the polymeric matrix during the production phase of synthetic fiber. The activity against microorganisms occurs through contact and/or diffusion. There are no antimicrobial agent disperses in activity through contact and show impact on the microorganism at the time of contact. In the event of diffusion, the antimicrobial agent reaches the outer environment away from the fiber surface, or polymer matrix, and shows activity on the microorganism [17–21].

A living germ, bacteria, or fungal has a cell wall of polysaccharides on the outermost surfaces. This structure ensures their integrity and protects them against the external environment. There is a semipermeable cell membrane on the cell wall. The cell wall and membrane stores, protects, and performs the cell's vital organelles, enzymes, genetic information, and transport. The type of activity of the antimicrobial agent against the microorganism is the main factor in its classification. If the antimicrobial agent only prevents the growth of the microorganism, it is called a biostatic effect; if it kills microorganisms, it is called biocidal effect [22–24].

Antimicrobial finishing processes have three different mechanisms [25]:

3.1 Controlled release

Most antimicrobial substances operate with a controlled oscillation mechanism. In this mechanism, the antimicrobial substance, which has already been applied to the textile material, is released at a certain speed in a controlled manner during use. This type of antimicrobial substance, which is removed when the textile material is washed, is very effective against microbes on or around the fiber surface. However, since it is constantly released during use, the amount of the textile material is gradually depleted at the end of the antimicrobial substance, and therefore the exhaustion process is depleted. On the other hand, the environmentally released antimicrobial substances are toxic to beneficial microorganisms and other creatures [24–26].

In recent years, studies have increased the use of silica carriers such as zeolite and microencapsulation technology for controlled oscillation in order to increase the strength of antimicrobial process or effect and cause less damage to the environment [25–27].

3.2 The regeneration principle

The renewal model was formulated by Gagliardi in 1962. This model, described in Gagliardi's article, is based on the application of a chemical finishing process product to fabrics that produce active germ killer (antiseptic) substances that are constantly renewed by adding bleaching substances during washing or exposure to ultraviolet light. This regeneration occurs when the covalent bonds in the chemically modified fiber are severed as a result of washing or photochemical effects, so that the model has an unlimited antimicrobial repository [27]. Although the regeneration technique has not yet been implemented, the microencapsulation technique is close to performing the function of this model. However, although the surface is suitable for a long period of time, microencapsulated antimicrobial substance storage is not unlimited [25, 27, 28].

3.3 Blocking or the blocking effect

The blocking or blocking mechanism for the protection of fabrics from microorganisms can be divided into two: (a) inert (ineffective) physical obstacle layers or coatings that are simply resistant to the passage of microorganisms into fabric or (b) layers or coatings with direct surface contact effect against microbial proliferation [27].

Fire, water, weather, and mildew resistant (FWWMR) end process is an example of obstacle coating. In this process, fabrics are coated with a mixture of organic and anorganic compounds containing fungicide. The blocking or blocking mechanism has been used to protect fabrics from mold yeast and decaying fungi with resin applications or chemical modification of cellulose with cyanoetylation or acetylation. When the finishing process containing flame-retardent agents and resins forms of finishing agent with covalent bonds, they are the most effective products against mold [27].

The product of the only antibacterial finishing process based directly on the concept of surface contact attachment obstruction is an organosilicon polymer containing hanging quaternary ammonium groups that form a biobarrier in the fabric [27].

Most of the antimicrobial agents used to manufacture commercial textiles have biocidal effects, but they show activity on microorganism in different ways [17]:

- They damage or inhibit the synthesis of the cell wall, which is critical for life and survival.
- They damage intracellular and non-cell matter transport by inhibiting cell membrane function.
- They cause the death of the microorganism by inhibiting the synthesis of the proteins that make up the building rocks of the cell and enzymes.
- By inhibiting nucleic acid (DNA and RNA) synthesis, they prevent the survival and proliferation of the cell.
- By inhibiting metabolic processes, they cause the death of the microorganism.

4. Antibacterial agents used in the textile industry

The most common antimicrobial substances used to give textile materials antimicrobial properties are quaternary ammonium compounds (QAC), polyhexamethylene biguanide (PHMB), chitosan, regenerated N-halamine compounds, peroxy acids, metal/metal salts, and triclosan. In addition, there are antimicrobial-enabled paints (e.g., metallic paints) that allow simultaneous dyeing and antimicrobial finishing processes [25]. The chapter is about metal-based antimicrobial finishing and triclosan-based antimicrobial finishing.

4.1 Metal-based antimicrobial finishing

Many heavy metals are toxic to microorganisms, both freely and in compounds, even at very low concentrations. Other heavy metals such as copper, zinc, and cobalt are also used in the production of antimicrobial textiles, but the most preferred are silver and silver compounds for this purpose [17, 29, 30]. In recent years, the nano-forms of metal and metal compounds have attracted attention as new generation biocides [30]. According to 2018 data, the most commonly used antimicrobial substances in the production of antimicrobial medical textiles are metal/metal salts (39.6%) [31]. The most commonly used metallic salts are silver, copper, zinc, and cobalt [31–33]. The global nano-silver market volume is estimated to exceed $3.3 billion in 2024 [34].

Metal and metal compounds cause oxidative stress in the microorganism, causing damage to microorganism lipid, protein, and DNA, resulting death [30]. The mechanism of action of the nano-forms of metal/metal compounds is similar. Silica such as zeolite, polymer matrixes, and various cross linking agents are used to stabilize nanoparticles in the structure, to provide controlled oscillation, and to ensure washing durability [30].

In synthetic fibers, metal and metal compounds can be added to the environment before fiber extraction or in the polymer stage before electrospinning and nano-fiber production. During its lifetime, metal ions are released causing biocidal effects in the presence of moisture. The amount of metal ion released varies depending on the chemical structure of the fiber, its surface feature, and the amount of metal/metal salt on the fiber [29].

The application of metals to natural fibers can only be done during the finishing process. Various strategies have been developed to improve binding and durability. Cotton was pre-treated with succinct acid anhydrites. Succinic acid anhydride acts as a ligand (atom, molecule, or ion attached to the central atom) for metal ions and provides very effective antibacterial activity by increasing the retention of metal salts (Ag^+ and Cu^{2+}) on the surface. In protein fibers (e.g., wool), aspartyl and glutamyl residues are thought to be binding groups for free carboxyl groups, most likely metal ions. Binding capacity can be further increased with EDTA with the ability to skip the tannin acid or metal ions that increase the serious restrictions due to technical and environmental problems; therefore, it is not accepted in commercial production [29].

4.1.1 Silver-based antimicrobial finishing

Silver has been used in many areas for centuries as a broad-spectrum antimicrobial substance with antibacterial, antifungal, and antiviral properties. Metallic silver, silver nitrate, and silver sulfadiazine forms have been used for many years to treat burns, wounds, and numerous bacterial infections [35]. Most metal ions are also known to have antimicrobial properties, but silver is best effective against bacteria, viruses, and other eukaryotic microorganisms [35]. Silver has very important advantages as an antibacterial substance. These benefits include the fact that silver is a very broad-spectrum antibiotic and has almost no bacterial resistance to silver, and there is no toxicity in low concentrations [35–37].

It is known that the use of silver in the treatment of burns and chronic ulcers in water disinfection dates back to the 1000 BC. In the literature, it is mentioned that silver was used as an eye drop in the 1800s, and then its used was reduced with the presence of penicillin, but 0.5% silver nitrate solution in the 1960s was widely used in burn treatment. In these years, silver's effectiveness against bacteria such as *Staphylococcus aureus*, *Pseudomonas aeruginosa*, and *Escherichia coli* has been proven. In 1968, silver sulfadiazine cream was obtained by combining silver nitrate with sulfonamide. This cream has been widely used in the treatment of burns due to its effectiveness against many microorganisms. The literature states that silver sulfadiazine is active against bacteria such as *E. coli*, *S. aureus*, *Klossiella* sp., and *Pseudomonas* sp. and also has antifungal and antiviral activities [35]. 1% silver nitrate solution is still used as eye antiseptic for various purposes in newborn babies [38]. Today, wound dresses containing different amounts of silver against antibiotic-resistant bacteria are used [35].

Concentrations greater than 0.5% are not generally preferred in silver solutions used for medical purposes. In these concentrations, silver allergy is not reported.

However, when using wound dress containing a high amount of silver ion in large wounds, a disease called argyrism can be found in the form of bluish and brown lesions in the skin and mucous membrane. This disease causes the removal of silver ions from the open wound for a long time [35, 39].

Metallic silver is actually inert, but when it comes into contact with the skin, the moisture and fluid of the wound on the skin make it ionized. Iodine silver is highly reactive. It connects to tissue proteins, causing structural changes in the bacterial cell wall and then the nuclear membrane, causing the death of the microorganism [35].

4.1.2 Mechanism and toxicity of silver

The mechanism of killing microorganisms by silver is still not very clear. The mechanism was attempted to be clearer by examining morphological and structural changes caused by metallic silver, silver ions, and silver nanoparticles in the bacterial cell. In light of the studies, it is known that silver is connected to the bacterial cell wall and cell membrane, interacting with thiol groups to inhibit respiratory enzymes, thus leading to the death of the microorganism [35, 36].

Liau and his colleagues studied the effect of silver ions on amino acids containing thiol (–SH) groups in 1997 [40].

A 2000 study by Feng and colleagues examined the morphological changes that silver ions have on gram-positive *S. aureus* and gram-negative *E. coli* bacteria. AgNO3 was used as an ion source in the study. Gram-positive *S. aureus* has been shown to be able to better resist silver ions due to its thick cell wall, which is typical of positive bacteria. Again, the study reported that DNA, which can only be copied while free, has become a more intense form within the cell, which shows that DNA has lost its ability to copy itself [41].

In his 2005 study, Holt and colleagues reported that the increase in the amount of potassium in the environment was detoxicated by the toxicity of silver against microorganisms [42].

Li and colleagues studied the antibacterial effect mechanism of silver nanoparticles on *E. coli* in a 2010 study. In this study, silver nanoparticles first disrupted the structure of the cell membrane and entered the cell and then inhibited the respiratory enzymes by relocating the hydrogen atoms (–S–Ag–) in the cysteine thiol (–SH) groups. The development and proliferation of bacteria stop if cell member permeability and respiratory of cell deteriorate [43].

Many studies are being conducted on the antimicrobial mechanism of nano-silver particles, but there is not enough work on toxicity. A limited number of studies conducted in in vitro conditions show that nano-silver particles are much more toxic than conventional silver and other heavy metals [35, 44]. Shapes, particle sizes, crystalline, surface properties, ambient humidity, ambient pH, cations in the environment, and their concentrations are among the particles that affect the toxicity of silver nanoparticles [45]. In vitro studies reveal that nano-silver particles cause damage to the brain, liver, and reproductive cells in mammals. In 1999, the FDA warned that the use of colloidal silver solutions containing micro- or nanoparticles could lead to neurological problems, headaches, skin irritation, weakness, stomach ailments, and kidney ailments. It is also reported that silver nanoparticles will affect rivers, lakes, and all living things that make up the ecosystem by blending into the food chain by mixing into the water. Washing machines produced in recent years, using nano-silver technology, are also objectionable in this context. In order to further clarify this issue, a large number of independent animal and clinical trials that are not supported by producers must be performed [35, 43, 46].

4.1.3 Silver contaminated waste and silver accumulation

Silver and its different forms are wide spectrum antibiotics. They have low risk of bacterial resistance, and their low concentrations are not toxic, and they have ease of application and low cost. Because of these advantages, silver and other forms of it are widely used in most areas and surfaces, which are being antimicrobial desired. It is also widely used in the production of antimicrobial textiles in different forms of Ag and silver (colloidal silver, silver salts, and elemental silver in powder form) [35–37].

Ag particles are applied to the textile surface using binder or cross-binding substances; it is possible to increase washing resistance. However, as a result of washing both during antimicrobial textile production and throughout its life cycle, most of the Ag particles on the textile surface mix into rivers, lakes, and groundwater along with wastewater, causing the accumulation of silver in the ecosystem. Disposable hygiene products are a similar situation [36]. Most antimicrobial textile products are released into washing water for 50% of the amount of silver at the end of three washings. And the textile products release 10–98% content of the silver into washing water at the end of 10 washings [47]. According to a study, up to 75% of silver may be released from textiles impregnated with Ag NPs in one washing cycle [48]. It is clear that silver accumulated in the ecosystem, water or soil, will have a toxic effect on all living organisms and reach the food chain [14, 35].

According to a study conducted in 64 countries on the release of silver from different products into nature, the United States is the country that releases the most silver into the environment, globally. The Asian continent is the continent which has the most silver emissions directly into the aquatic environment and land [49]. According to a report, 68% of the global silver consumption is used for water treatment and 32% for other uses. And 3.4–40 metric tons of silver are used in textiles per year [5]. In the United States, 29% of the silver used in different industries is released into the aquatic environment, and 69% are known to be dumped in solid waste storage [50]. In recent years nano-silver consumption in textiles like other industries has been increasing rapidly also [51]. The regions where antimicrobial medical textiles containing metallic salts such as copper, zinc, cobalt, mainly silver most used are North America (39% of market volume), Europe (23% of market volume), the Asia Pacific regions (30% of market volume) and the rest of the world (7% of the market volume) respectively [31, 48]. The highest use rate belongs to North America because hospital infection and cardiovascular disease rates are high in this region [31].

4.2 Triclosan-based antimicrobial finishing

Triclosan has been widely used in commercial products for many years as an antimicrobial substance used in soaps, deodorants, cosmetics, cleaning lotions, plastics, toothpastes, and antibacterial textiles [52–56]. The European Union's consumption of triclosan in 2006 is reported to be approximately 450 tons. It is reported that 85% of this is used in personal care products, 5% in textile products, and 10% in plastics and products that come into contact with food [54, 57]. Triclosan is also frequently used in the textile industry. Triclosan is used to prevent the formation of bad odor in wool; to prevent the reproduction of bacteria and fungi in synthetic, mixtures, and non-woven textile materials; and to keep mites away from textile materials [57].

75–210 metric tons of triclosan are used in textiles per year globally [5]. According to a 2009 report by the Australian government, between 2001 and 2005, the amount of triclosan contained in textile products exported to Australia varied between 1 and 20%. The report stated that between 2001 and 2005, textile products containing approximately 1 ton of triclosan were used. In the same report, it is stated that triclosan in used in Australia in wool bed-duvet production, upholstery fabrics, towels, woolly textile products, preparatory fabric production, marine and sports clothes, socks, underwear, shoe linings, zippers, gloves, surgical masks, non-woven products, sleeping bags, and insulation textiles [57]. Triclosan can be added to the textile materials during the fiber production stage and can be applied as a finishing process or transferred in the form of coating [57].

4.2.1 Mechanism and toxicity of triclosan

Triclosan is known to have bacteriologic effects on gram-positive and gram-negative bacteria, as well as antifungal and antiviral properties [53, 56]. Triclosan inhibits lipid synthesis by blocking enoyl-acyl reductase (ENR) of the microorganism. Thus, it prevents the development of the microorganism and its proliferation of division [53].

In 1986 in accordance with the European Union Cosmetics Directive, triclosan has been confirmed that it can be used in materials in contact with foods of up to 0.3% concentration as protective material, 5 mg/kg textile materials (especially in sportswear), and 0.3% concentration of plastic (plastic packaging, brushes) materials [54]. The Japanese government has stated that in cosmetics, the maximum amount of triclosan that can be used is 0.1%. In oral care in Canada, the amount of triclosan allowed in their products is 0.03%, and in cosmetic products it is 0.3%. According to a 2009 report by the Australian government, with regard to triclosan, eyes, respiratory system, and skin have been described as being irritating and toxic to inhalation [57].

Studies on the effects of triclosan on human health are usually carried out with mice, rabbits, dogs, and monkeys [53, 54]. Triclosan is taken into the body through the skin, nose, and mouth during contact with products containing triclosan. In addition, triclosan has contaminated the sea, lake, and groundwater and has reached the food chain, especially from foods such as seafood; triclosan enters the human body [53]. A study of 36 breastfeeding mothers who stated that they used personal care products containing triclosan as a result of a series of studies in America found triclosan in the mothers' milks [53]. Studies have shown that triclosan affects androgens in the male body and estrogen in the female body. Triclosan was found to affect the transport between the fetus and the placenta in the bodies of pregnant sheep, which has been reported that this can cause abnormal development. It has also been reported that triclosan can trigger breast cancers, especially in females. A number of studies on rabbits have been reported to reduce the sperm count in male rabbits and cause tissue destruction in reproductive organs, disrupting masculinity hormones [53].

The thyroid is known to have vital effects on development and metabolism. The thyroid hormone is a highly effective hormone in the development of fetuses and young children. Studies have shown that triclosan lowers thyroid hormone levels in rabbits and changes metamorphosis time in frogs [53, 58].

4.2.2 Triclosan contaminated waste and triclosan accumulation

Water supplies all over the world have been contaminated with triclosan due to wide commercial use in commercial products. In a 1999–2000 study conducted in the United States, samples from different water sources were examined in terms of 95 different chemicals, and as a result, one of the chemicals with the highest concentration was triclosan. Again, the researchers found a very high amount of triclosan in the bodies of marine creatures in particular. The Environmental Protection Agency reported that some of the triclosan in the environment was disrupted by the effect of ultraviolet rays and turned into toxic dioxins. It is reported that the access of dioxins to the food chain will have bad consequences [52]. Because the demolition products of triclosan are also toxic [59]. Again, the formation of cancer is associated with triclosan exposure [59]. According to a study, antimicrobial textile products containing triclosan are sold in 64–84% of the triclosan wash water at the end of 10 washings [57].

5. Bioactive plant-based environment-friendly antimicrobial finishing

Bio-functionalization of textiles with natural bioactive agents with antimicrobial properties is becoming increasingly important because they are not toxic, skin, and environment-friendly. These antimicrobial compounds extracted from most plants are phenols, polyphenols (simple phenols, phenolic acids, quinines, flavonoids, tannin, coumarin, etc.), terpenoids, essential oils, alcoholicoids, lectins polypeptides, and polyacethylenes. Most of these substances obtained from plants are colorful and are natural antimicrobial dyes and pigments used for the dyeing of both natural and synthetic fibers [30, 60–65]. Eco-friendly pigments can be obtained with fermentation of bacteria and fungi [30, 66, 67]. Different methods are mentioned in the literature to increase washing habits of bioactive vegetable-based antimicrobial compounds uncinated on textile fiber: resin application with cross-binding agent, glyoxal, and glycol [30, 68]; sol-gel matrix of liquid bioactive compounds, such as essential oils [30, 69]; and application with microcaps or with the pad-dry-cure method [30, 70–72].

Hydrogen peroxide is a natural antimicrobial produced against invasive bacteria in human cells. It is also found in honey as a preservative. Antimicrobial activity of hydrogen peroxide against bacteria, mold, fungi, algae, and viruses is known. The finishing processes and substances with hydrogen peroxide have become popular and commercialized in recent years [14].

It is thought that the importance of antimicrobial-effective herbal (such as vegetable wastes etc.) and animal-derived natural materials will increase for reducing the waste load (production, during its lifetime, and at the end of its lifespan) and engaging in more environment-friendly manufacturing [14].

6. Conclusion

In today's world, the role of the textile industry is very important. While the textile industry initially met traditional human needs such as dressing with yarn and fabric production and home textiles, today due to rising living standards, textiles have become much more technological and functional with diversified human requirements. It is also an important industry sector for both countries in the growth and development process (rather than traditional textile production) and countries that have completed their development (rather than high technological textile production). However, despite all these advantages, the textile industry causes a large amount of waste and environmental pollution.

At different stages of textile production, numerous chemicals and auxiliary substances are used, many of which are toxic and harmful to the environment and human beings. As a result of these production stages, a large amount of solid, liquid, gas, and sludge form waste is exposed and causes pollution. Noise pollution is also another negative result of the textile industry. Textile finishing operations are the processes where high amounts of water are used, so high amounts of wastewater (with high chemical load) occur. Therefore, the biggest problem of the textile industry is this wastewater burden. According to some studies, 20% of all fresh water pollution is made by textile treatment and dyeing [73]. Textile wastewater needs to be properly purified to reduce environmental damage. In this context, the selection of chemicals and dyes with less environmental damage or environment-friendly finishing operations is also important in this context.

Any textile product has been subjected to washing, dry cleaning, and ironing many times during its service life. With each wash, the active chemical finishing agent in its structure is mixed into washing water, which then threatens the entire ecosystem by mixing into the sea, lakes, and underground waters, and is consequently used by water and soil plants contaminated with antimicrobial lice chemicals to be included in the food chain. Again, the seas and rivers contaminated with antimicrobial substances threaten water creatures and the human health as a result of consuming these creatures.

Studies on antimicrobial textiles have focused mainly on the synthesis of antimicrobial matter and its performance against microorganisms and washing durability. However, the effects of waste/wastewater content on the user's skin and health and all other creatures through contact/respiratory/consumption are needed to be further studies during the production of antimicrobial textiles, during and at the end of its lifecycle [74]. Antimicrobial agents derived from natural sources are safe for human and the environment, but the spectrum of activity and efficiency is not as good as the synthetic ones. To achieve this, more research work is needed in the field. Hence, natural antimicrobial agents derived from plant sources would be of prime importance in the future [75]. It is so urgent to protect and conserve the natural ecosystem of the earth, thereby restoring the global sustainability.

Author details

Candan Akca
Candan Akca Textile Technology Department, Salihli Vocational School, Manisa Celal Bayar University, Manisa, Turkey

*Address all correspondence to: candanakca@yahoo.com; candan.can@cbu.edu.tr

References

[1] Sivaram M, Gopal M, Barik D. Toxic waste from textile industries. In: Energy from Toxic Organic Waste for Heat and Power Generation. UK: Elsevier; 2019. pp. 43-54. DOI: 10.1016/B978-0-08-102528-4.00004-3

[2] Global Textile Market [Internet]. 2020. Available from: https://www.globenewswire.com/newsrelease/2019/05/02/1814045/0/en/Global-Textile-Market-Will-Reach-USD-1-207-Billion-By-2025-Zion-Market-Research.html [Accessed: 24 January 2020]

[3] Afraz N, Uddin F, Syed U, Mahmood A. Antimicrobial finishes for textiles. Current Trends in Fashion Technology and Textile Engineering. 2019;**4**(5):87-94. DOI: 10.19080/CTFTTE.2019.04.555646

[4] Gao Y, Cranston R. Recent advances in antimicrobial treatments of textiles. Textile Research Journal. 2008;**78**(1):60-72. DOI: 10.1177/0040517507082332

[5] Windler L, Height M, Nowack B. Comparative evaluation of antimicrobials for textile applications. Environment International. 2013;**53**(C):62-73. DOI: 10.1016/j.envint.2012.12.010

[6] Global Antimicrobial Textile Markets [Internet]. 2020. Available from: https://www.globenewswire.com/news-release/2019/11/08/1943912/0/en/Global-Antimicrobial-Textile-Markets-Forecast-to-2024-Synthetic-Organic-Compounds-Active-Agent-Expected-to-Drive-the-Market.html [Accessed: 24 January 2020]

[7] Global Functional Textile Finishing Agents Market [Internet]. 2020. Available from: https://www.grandviewresearch.com/industry-analysis/functional-textile-finishing-agents-market [Accessed: 24 January 2020]

[8] Wet Tissue and Wipe Market Industry Analysis [Internet]. 2020. Available from: https://www.technavio.com/report/wet-tissue-and-wipe-market-industry-analysis [Accessed: 24 January 2020]

[9] Global Baby Diaper Market Size [Internet]. 2020. Available from: https://www.globenewswire.com/news-release/2019/07/11/1881415/0/en/Global-Baby-Diapers-Market-Size-Will-Reach-USD-67-46-Billion-By-2022-Zion-Market-Research.html [Accessed: 24 January 2020]

[10] Yalçın İ, Küçükali M, Sezgin H. Risks and management of textile waste. In: Gothandam K, editor. Nanoscience and Biotechnology for Environmental Applications, Environmental Chemistry for a Sustainable World Book 22. Switzerland: Springer Nature; 2019. pp. 29-53. DOI: 10.1007/978-3-319-97922-9_2

[11] Körlü A. Use of Ozone in the Textile Industry. Rijeka: Intechopen; 2018. DOI: 10.5772/intechopen.81774

[12] Can Y, Akaydın Y. Effects of laundering process on pilling characteristics of the cotton plain fabric. Pamukkale Üniversitesi Mühendislik Bilimleri Dergisi. 2013;**19**(4):170-173. DOI: 10.5505/pajes.2013.63935

[13] Detergents Effects on Marine Life [Internet]. 2020. Available from:https://www.lenntech.com/aquatic/detergents.htm [Accessed: 24 January 2020]

[14] Uddin F. Environmental concerns in antimicrobial finishing of textiles. International Journal of Textile Sciences. 2014;**3**(1A):15-20. DOI: 10.5923/s.textile.201401.03

[15] Domina T, Koch K. The textile waste lifecycle. Clothing and Textiles Research Journal. 1997;**26**(3):505-515. DOI: 10.1177/0887302X9701500204

[16] Toxic and Hazardous waste [Internet]. 2020. Available from: https://www.epa.gov./environmental-topics/land-waste-and-cleanup-topics [Accessed: 24 January 2020]

[17] Morais DS, Guedes RM, Lopes MA. Antimicrobial approaches for textiles: From research to market. Materials. 2016;**9**(6):498. DOI: 10.3390/ma9060498

[18] Shahidi S, Wiener J. Antibacterial Agents in Textile Industry. Rijeka: Intechopen; 2012. pp. 387-406. DOI: 10.5772/46246

[19] Korkmaz BR. Günümüz Tekstillerinde Kumaşa Kazandırılan Teknik Özellikler [Yüksek Lisans Tezi]. Marmara Üniversitesi, Güzel Sanatlar Enstitüsü: Çiğdem Çini; 2015

[20] Bshena O, Heunis TD, Dicks LM, Klumperman B. Antimicrobial fibers: Therapeutic possibilities and recent advances. Future Medicinal Chemistry. 2011;**49**(10):691-707. DOI: 10.4155/fmc.11.131

[21] Shishoo R. Plasma Technologies for Textile. Cambridge, UK: Elsevier; 2007. ISBN-10: 1-84569-257-8

[22] Bacterial Resistance [Internet]. 2020. Available from: https://amrls.cvm.msu.edu/microbiology/bacterial-resistance-strategies/introduction/introduction [Accessed: 24 January 2020]

[23] Glazer AN, Nikaido H. Microbial Biotechnology: Fundamentals of Applied Microbiology. Cambridge, MA, USA: Cambridge University Press; 2007. ISBN-10: 0-511-34136-9

[24] Rahman MA, Ahsan T, Islam S. Antibacterial and antifungal properties of the methanol extract from the stem of *Argyreia argentea*. Bangladesh Journal of Pharmacology. 2010:41-44. DOI: 10.3329/bjp.v5i1.4700

[25] Arslan P, Tayyar AE. Tekstil Alanında Kullanılan Antimikrobiyal Maddeler, Çalışma Mekanizmaları, Uygulamaları ve Antimikrobiyal Etkinlik Değerlendirme Yöntemleri. Düzce Üniversitesi Bilim ve Teknoloji Dergisi. 2016;(4):935-966. ISSN: 2148-2446

[26] Schindler WD, Hauser PJ. Chemical Finishing of Textiles. UK: Woodhead Publishing: 2004. ISBN: 9781845690373

[27] Vigo TL. Textile Processing and Properties Preparation, Dyeing, Finishing, and Performance. ABD: Elsevier; 1983. ISBN: 0 444 88224 3

[28] Gagliardi DD. Antimicrobial Finishes. American Dyestuff Reporter. 1962;**51**(1):49

[29] Bulan J, Celikiz G. Use of heavy metals in textile wet processing. Textile Chemist and Colorist. 1990;**22**(1):23-27

[30] Simoncic B, Tomsic B. Structures of novel antimicrobial agents for textiles—A review. Textile Research Journal. 2010;**80**(16):1721-1737. DOI: 10.1177/0040517510363193

[31] Antimicrobial Medical Textiles Market [Internet]. 2020. Available from: https://www.reportsanddata.com/report-detail/antimicrobial-medical-textiles-market [Accessed: 24 January 2020]

[32] The market for antimicrobial additives/biocides in plastics and textiles worldwide 2011- 2025: Development, Strategies, Markets, Companies, Trends, Nanotechnology, Home page [Internet]. 2020. Available from: http://www.hkc22.com/antimicrobials.html [Accessed: 24 January 2020]

[33] Giannossa LC, Longano D, Ditaranto N. Metal nanoantimicrobials for textile applications. Nanotechnology Review. 2013;**2**(3):241. DOI: 10.1515/ntrev-2013-0004

[34] Nanosilver Market [Internet]. 2020. Available from: https://www.gminsights.com/pressrelease/nanosilver-market [Accessed: 24 January 2020]

[35] Rai M, Yadav A, Gade A. Silver nanoparticles as a new generation of antimicrobials. Biotechnology Advances. 2009;**27**(1):76-83. DOI: 10.1016/j.biotechadv.2008.09.002

[36] Bozoğlu A. Türkiye'de Çıkan Doğal Zeolitin Bebek Bezlerinde Antimikrobiyal Madde olarak Kullanılması [thesis]. İstanbul Teknik Üniversitesi; 2010

[37] Silver Properties [Internet]. 2020. Available from: http://www.beggandco.com/admin/downloads/IMSL%20White%20Paper-JIS%20Z%202801-1005.pdf [Accessed: 24 January 2020]

[38] Çetin ET. Dezenfeksiyon, Antisepsi, Sterilizasyon. Turkey: İstanbul Tıp Fakültesi Yayını; 1982

[39] Grönvall H. On argyrosis and concretion in the Lacrymal Sac. Acta Ophthalmologica. 2009;**21**(4):17

[40] Liau SY, Read DC, Pugh WJ, Furrn JR, Russell AD. Interaction of silver nitrate with readily identifiable groups: Relationship to the antibacterial action of silver ions. Letters in Applied Microbiology. 1997;**25**(4):279-283. DOI: 10.1046/j.1472-765x.1997.00219.x

[41] Feng QL, Wu J, Chen GQ, Cui FZ, Kim T, Kim JO. A Mechanistic Study of the Antibacterial Effect of Silver Ions on *Escherichia coli* and *Staphylococcus aureus*. ABD: John Wiley and Sons; 2010. pp. 662-668. DOI: 10.1002/1097-4636(20001215)52:4<662::aid-jbm10>3.0.co;2-3

[42] Holt K, Bard A. Interaction of silver (I) ions with the respiratory chain of *Esherichia coli*: An electrochemical and scanning electrochemical microscopy study of the antimicrobial mechanism of micromolar Ag. Biochemistry. 2005;**44**(39):13214-13223. DOI: 10.1021/bi0508542

[43] Li W, Xi X, Shi Q, Zeng H, Ou S, Chen Y. Antibacterial activity and mechanism of silver nanoparticles on *Escherichia coli*. Applied Microbial and Cell Physiology. 2010;**85**(4):1115-1122. DOI: 10.1007/s00253-009-2159-5

[44] Nano-silver Report [Internet]. 2020. Available from: http://www.foeeurope.org/activities/nanotechnology/Documents/FoE_Nanosilver_report.pdf [Accessed: 24 January 2020]

[45] Jones C, Hoek E. A review of the antibacterials effects of silver nanomaterials and potential implications for human health and the environment. Journal of Nanoparticules Research. 2010;**12**(5):1531-1551. DOI: 10.1007/s11051-010-9900-y

[46] ABG L. Silver I: It's antibacterial properties and mechanism of action. Journal of Wound Care. 2002;**11**(4):125-130. DOI: 10.12968/jowc.2002.11.4.26389

[47] Silver Accumulation [Internet]. 2020. Available from: https://www.kemi.se/global/pm/2012/pm-1-12-antibact-en.pdf [Accessed: 24 January 2020]

[48] Rosenberg M, Ilic K, Juganson K, Ivask A. Potential ecotoxicological effects of antimicrobial surface coatings: A literature survey backed up by analysis of market reports. PeerJ. 2019;(1):2-34. DOI: 10.7717/peerj.6315

[49] Eckelman MJ, Graedel TE. Silver emissions and their environmental impacts: A multilevel assessment. American Chemical Society. 2007;**41**(17):6283-6289. DOI: 10.1021/es062970d

[50] Silver Accumulation in ABD [Internet]. 2020. Available from:

https://www.who.int/ipcs/publications/cicad/en/cicad44.pdf [Accessed: 24 January 2020]

[51] Lombi E, Donner E, Scheckel K, Sekine R. Silver speciation and release in commercial antimicrobial textiles as influenced by washing. Chemosphere. 2014;(111):352-358. DOI: 10.1016/j.chemosphere.2014.03.116

[52] Triclosan [Internet]. 2020. Available from: www.minddisrupted.org [Accessed: 24 January 2020]

[53] McGinnis D. Toxicological Profile of Triclosan in the Aquatic Environment. Department of Marine and Environmental Systems Florida Institude of Technology Melbourne; 2008

[54] European Commission, Health and Consumer Protection Directorate-General, Scientific Committee on Consumer Products Opinion on Tirclosan [Internet]. 2009. Available from: https://ec.europa.eu/health/sites/health/files/scientific_committees/consumer_safety/docs/sccs_o_023_note.pdf [Accessed: 24 January 2020]

[55] Minnesota Department of Health, Drinking Water Contaminants of Emerging Concern Program, Triclosan Exposure and Toxicity Summary. 2010

[56] Orhan M, Kut D, Güneşoğlu C. Use of triclosan as antibacterial agent in textiles. Indian Journal of Fibre and Textile Research. 2007;**32**(1):114-118

[57] Australian Government Department of Health and Ageing NICNAS, Priority Existing Chemical Assessment Report Triclosan [Internet]. 2009. Available from: https://www.nicnas.gov.au/chemical-information/factsheets/chemical-name/triclosan [Accessed: 24 January 2020]

[58] Triclosan Toxicity [Internet]. 2020. Available from: http://www.beyondpesricides.org/pesticides/factsheets/Triclosan%20cited.pdf [Accessed: 24 January 2020]

[59] Weatherly LM, Gosse JA. Triclosan exposure, transformation, and human health effects. Journal of Toxicology and Environmental Health, Part B: Critical Reviews. 2017;**20**(8):447-469. DOI: 10.1080/10937404.2017.1399306

[60] Cowan MM. Plant products as antimicrobial agents. Clinical Microbiology Reviews. 1999;**12**(4):564-582. DOI: 10.1128/CMR.12.4.564

[61] Holme I. Finishes for Protective and Military Textiles [Internet]. 2005. Available from: https://www.tib.eu/en/search/id/tema%3ATEMA20050605954/Finishes-for-protective-and-military-textiles/ [Accessed: 24 January 2020]

[62] Son YA, Kim BS, Ravikumar K, Kim TK. Berberine finishing for developing antimicrobial nylon 66 fibers: Exhaustion, colorimetric analysis, antimicrobial study, and empirical modeling. Journal of Applied Polymer Science. 2007;(103):1175-1182. DOI: 10.1002/app.25364

[63] Gupta D, Khare SK, Laha A. Antimicrobial properties of natural dyes against gram-negative bacteria. Coloration Technology. 2004;**120**(4):167-171. DOI: 10.1111/j.1478-4408.2004.tb00224.x

[64] Han S, Yang Y. Antimicrobial activity of wool fabric treated with curcumin. Dyes and Pigments. 2005;**64**(2):157-161. DOI: 10.1016/j.dyepig.2004.05.008

[65] Giri Dev VR, Venugopal J, Sudha S, Deepika G, Ramakrishna S. Dyeing and antimicrobial characteristics of chitosan treated wool fabrics with henna dye. Carbohydrate Polymers. 2009;**53**(2):646-650. DOI: 10.1016/j.carbpol.2008.09.003

[66] Hobson DK, Wales DS. Green colorants. Journal of the Society of Dyers and Colourists. 1998;**114**(2):42-44. DOI: 10.1111/j.1478-4408.1998.tb01944.x

[67] Alihosseini F, Ju KS, Lango J, Hammock BD, Sun G. Antibacterial colorants: Characterization of prodiginines and their applications on textile materials. Biotechnology Progress. 2008;**24**(3):742-747. DOI: 10.1021/bp070481r

[68] Joshi M, Ali SW, Rajendran S. Antibacterial finishing of polyester/cotton blend fabrics using neem (*Azadirachta indica*): A natural bioactive agent. Journal of Applied Polymer Science. 2007;**24**(3):793-800. DOI: 10.1002/app.26323

[69] Haufe H, Muschter K, Siegert J, Böttcher H. Bioactive textiles by sol–gel immobilized natural active agents. Journal of Sol–Gel Science and Technology. 2008;**45**(1):97-101. DOI: 10.1007/s10971-007-1636-5

[70] Boh B, Knez E. Microencapsulation of essential oils and phase change materials for applications in textile products. Indian Journal of Fibre and Textile Research. 2006;**31**(1):72-82

[71] Thilagavathi G, Bala SK, Kannaian T. Microencapsulation of herbal extracts for microbial resistance in healthcare textiles. Indian Journal of Fibre and Textile Research. 2007;**32**(3):351-354

[72] Islam S. Butola BS. Nanomaterials in the Wet Processing of Textiles. ABD: Scrivener Publishing; 2018. ISBN: 978-1-119-45984-2

[73] Textile Industry and Environment [Internet]. 2020. Available from: https://www.greenofchange.com/textile-pollution [Accessed: 17 February 2020]

[74] Rovira J, Domingo J. Human health risks due to exposure to inorganic and organic chemicals from textiles: A review. Environmental Research. 2019;**168**(1):62-69. DOI: 10.106/j.envres.2018.09.027

[75] Reshma A, Priyadarisini B, Amutha K. Sustainabel antibacterial finishing of fabrics using natural bioactive agents—A review. International Journal of Life Science and Pharma Research. 2018;**8**(4):10-20. DOI: 10.22376/ijpbs/lpr.2018.8.4.L10-20

Chapter 3

Textile Wastes: Status and Perspectives

Burçin Ütebay, Pinar Çelik and Ahmet Çay

Abstract

The world population has grown tremendously in the past few decades, and the same period also witnessed improvements in living standards in general. These two developments have augmented the consumption of textiles, which in turn increased textile production. Global production of all apparel and textile fibers amounted to more than 110 million tons annually, leading to the generation of high amount of textile wastes. In order to ensure sustainability and reduce environmental impacts in the textile and apparel sector, utilizing a circular economy model is of utmost importance. Recycling of textile waste is a requirement for the implementation of a circular model. This study presents a general evaluation of management of textile wastes, in terms of ensuring sustainability and minimizing environmental impacts.

Keywords: textile wastes, sustainability, waste management, recycling, circular economy

1. Introduction

Textile is an ancient industry that goes back to the beginning of the history of humanity, and its products range from products of daily usage to technical ones. All kinds of garments obtained by processing knitted, woven and nonwoven fabrics are categorized as the apparel sector. Actually, textile production is one of the main industries that affect global environmental pollution, as both the production and the processing of the necessary raw materials are contributing factors to pollution. Another important aspect of the problem is the waste that results from both production and consumption of the textile goods. Though technically all waste in the textile and garment sector can be recycled, unfortunately, only a small amount is recycled. As long as the linear system currently utilized in the production goes on, it seems that we will not be able to use the resources efficiently and reduce the environmental pollution.

Given that the current global trends persist, by 2050, the textile sector is expected to represent a quarter of the world carbon budget—26%, to be precise. The figures are colossal: If the current trends do not shift, the textile and apparel sector's nonrenewable raw material usage will reach 300 million tons and the amount of microplastic released to the oceans will reach 22 million by 2050 [1].

One key concept to analyze and understand the situation is the linear economy. As the dominant model of production at least since the Industrial Revolution, linear economy basically works as "subtract the raw material from the source, convert it into a product, sell the product to the consumer, which eventually gets disposed of by the consumer after usage." Under this model, products discarded by the consumer become waste and are generally disposed of ending up in landfills or by incineration [2].

The basis of the linear economy approach is the consumption of the raw material required for production. It seems that the limited resources available to us in the world will not provide the conditions for the current dominant economic model to go on as today. The actual perception of raw materials is not sustainable. Moreover, linear economy-oriented production and business models become a burden for the environment—the environmental aspect, the damages they cause and the waste that results from them are generally not considered.

An alternative to this traditional production model is what is called a "circular economy": "A circular economy is based on the principles of designing out waste and pollution, keeping products and materials in use, and regenerating natural systems" [3].

In contrast to the negative aspects of the linear economic approach, the circular economy is seen as a sustainable development model for the future and increasingly stands out as an approach that is becoming widespread in the world. The circular economy is a systems model in which every part of a product is considered as a valuable resource that should be handled with care and resources are circulated again and again through closed loops.

The basis of the circular economy model is to expand the life-span of a product via repair, reuse, remanufacturing and recycling, so that resources are used more efficiently and the need for new products and virgin raw material is reduced or ideally eliminated [2].

In order to ensure sustainability and reduce environmental impacts in the textile and apparel sector, utilizing a circular economy model is of utmost importance. Recycling of textile waste is a requirement for the implementation of a circular model. This study presents a general evaluation of recycling of textile wastes, in terms of ensuring sustainability and minimizing environmental impacts.

2. Fashion and sustainability

Decisions of consumption have obtained a significant role in today's world—the choice of what you eat, where you go, what you wear and so on has become decisive factors of showing who you are and what you want to be, of displaying one's social identity. When one talks about consumer behavior in the textile sector, fashion is the key concept. Fashion presents the markers for social differentiation, mobility and identity, and allows a person to present one's identity—moreover, this dynamism and speed of fashion allow people to change their conceived identity [4–6].

Certain historical, social and cultural changes can be said to have shaped contemporary fashion. The first big leap was the Industrial Revolution—the possibility of producing *en masse* made it possible for the consumers to pick their desired products from a much wider range at lower prices, which, in turn, meant that a wider portion of the society could afford to have the pleasure of purchasing the desired products. In other words, "...fashion, which had been the epitome of luxury, was democratized and consumption behavior began to change" [7]. Industrialization also triggered changes in the economic structures: Autarky lost ground and urbanization has become a widespread phenomenon. The fact that an increasing number of people started to move into cities enhanced the commercial activities. Starting with the 1900s, new types of trade companies and retail shops emerged in the cities. Consumption, which was mainly a subject of certain periodical trade events, has become a fundamental part of the daily routine of the people. The new shops enabled people to buy clothes any given day. This new dynamism of fashion provided retailers and marketers with new commercial opportunities, as "being fashionable" became an important aspect of consumer behavior [7, 8].

Although the modern concept of "consumer culture" goes as far back as three centuries, the last few decades have witnessed unprecedented growth of purchasing and disposal of textile goods. The emergence of fashion as a defining force in consumer behavior resulted in overconsumption. The logic of fashion constantly pushes forward the need for change, which results in generating more and more requests to supply newer, fresher and more contemporary goods [9, 10]. Until the 1990s, the general tendency of fashion retailers was to release two main collections—spring/summer and autumn/winter—each year; however, during the 1990s, drastic changes occurred: The so-called era of "super cheap and super fast" arrived [11]. The increasing ability to outsource production to low-cost regions of a globalized world and thus to produce much cheaper clothes, combined with the inherent dynamism of fashion, paved the way for the emergence of what we nowadays conceive as the "modern fashion business" [12].

This fundamental role of dynamism inevitably positioned time as a crucial factor for the competitiveness of fashion companies. Consumers are conditioned to expect newness; thus, brand new products need to arrive at the stores with short time intervals constantly. This objective is accompanied by limited ranges and rapid stock turnaround for the companies [13].

The dominance of fast fashion and just-in-time production in the textile industry has led to more frequent seasons and minicollections in-between seasons, which has led to the arrival of new cheap items to the stores every week, even, in some cases, every day. It is a chain reaction: increase in the creation of new fashion trends, desires to experience the new spurring out of control, consumers buying more and more, and eventually overconsumption. This new concept of seasonal new collection brings about more incentives of buying for the consumer and, thus, increases the rate of textile consumption. However, the fashion industry not only has an impact on people but also has a big impact on the environment [10, 14].

The concerns about the environment are rapidly growing in today's world and are shared by the fashion firms and the consumers. Textile production is an important source of human-made adverse impact on the environment, as the sector uses huge amounts of water, pesticides and chemicals. Attempts to establish guidelines for sustainability in the production phase, such as ISO 14000, are a reflection of this fact, and this aspect is quite relevant for fashion firms. On the other hand, the consumers are getting more and more conscious about the social and environmental problems, which have a direct effect on the consumption choices of the consumers, as in eco-fashion consumption [15–17].

Even though this burning issue is gaining more importance in all sectors, it can be said that textile lags behind other sectors, for example the industrial design, in terms of research and development about modes of production that would be more efficient for the conservation of the environment or ways to get the consumers more engaged in topics of sustainability. The industry needs more innovation in the aspects of design, manufacture, consumption and business within a sustainable framework [18, 19].

Sustainability is indeed a burning issue, and the following data demonstrate how important it is for the world to achieve greater success in the textile sector. Textile production, a sector that goes back to ancient times and has always maintained its pivotal role in human life, still has a paramount place in industry if one takes a look at the global output and employment numbers [20]. According to the Zion Market Research's report, the textile market was approximately valued at USD 858 billion in 2018 globally and is estimated to generate around USD 1207 billion by the year 2025, at a CAGR of around 5% between 2019 and 2025 [21]. The global garment and textile industries employ 60 million to 75 million people worldwide [22]. The total volume of the production of the sector around the world is expected to exceed 99 million tons annually. These numbers are evidence to the importance of the applications of the industry for environment [23].

Clothing is an essential human need, and the textile and clothing industry delivers goods to satisfy this basic necessity. But this vital sector presents serious social and ecological problems in many instances of the supply chain—from fiber production, spinning, fabric production, dyeing and finishing, to clothing production [24, 25]. However, the increasing price pressure over fashion companies in the last decades does not help the companies in developing more sustainable production models. The price pressure has led many textile companies to outsource their production, which caused the bulk of the European and US clothing production shift to developing economies in Asia. The part of the value creation chain that remains in the Western countries is mostly limited to value-added services such as design and overall brand management. This production shift, with the relocation of a big part of the value chain in lower labor cost countries, presents a new challenge for sustainability, as the surveillance and control over labor and ecological practices in the production sites of the supply chain have become much more complicated [25, 26].

Sustainability, a word more frequently used every day nowadays, may sound very familiar, but it is difficult to define, understand and adopt in industrial practices. An apt definition for the term sustainable development, coined by Brundtland (formerly the World Commission on Environment and Development), is as follows: "The development that meets the needs of the present without compromising the ability of future generations to meet their own needs" [27]. Below is a list of the main obstacles for an environmentally sustainable textile and apparel sector:

2.1 Consumption of water

Studies show that, in terms of consumption and pollution of clean water, textile and related industries are only surpassed by agriculture [28]. The sheer amount of water used during textile production, especially wet processing, tells a lot: to process a kilogram of fabric, 80 to 150 liters of water is used, along with other chemicals [29]. About 4% of global freshwater withdrawal, which corresponds to 93 billion cubic meters of water, is utilized annually by the textile sector, if cotton farming is included. Clothing is responsible for more than 60% of this amount [1].

2.2 Global warming

The average temperature on Earth is constantly increasing, but especially since the Industrial Revolution, the rate at which the average temperature on Earth has been rising is too rapid—the phenomenon known as global warming. Various estimates put the rise at 0.6–0.8°C, which corresponds to a rise 10 times faster than the calculated normal. This man-made global warming is due to the amount of greenhouse gases released, such as carbon dioxide and chlorofluorocarbon, of the use of fossil fuels as well as of other developments [30].

Textile and apparel production has a major role in this global phenomenon. In 2015, greenhouse gas (GHG) emissions from textile production were responsible for 1.2 billion tons of CO_2 equivalent of greenhouse gas (GHG). This figure exceeds the emissions that result from all international flights and maritime shipping combined—two fundamental means of transportation [1]. Transferring final products produced in developing countries to the shops in the developed ones necessitates long-distance maritime transportation, which further increases the total consumption of nonrenewable fuel [31].

Given that the actual trends of the sector do not alter, by 2050, 26% of the carbon budget and 300 million tons of crude oil will be consumed by the textile industry—a significant change compared with 2% and 98 million tons, respectively, in 2015 [1].

The textile industry utilizes much energy with little efficiency. The chemical processing leg of the production mostly utilizes thermal energy to heat water and dry fabrics, while spinning and weaving legs of the production mostly utilize electrical power [32, 33]. The consumption of electricity to produce 60 billion kilograms of fabrics worldwide per year is calculated to be nearly 1 trillion kilowatt hours [30].

Textile manufacture is also a source of NOx and SOx emissions, solvent release during drying of coatings or cleaning operations and volatile organic compounds (VOCs) [31].

2.3 Environmental pollution

About a quarter of chemicals produced globally are used in the textile industry [34]. Numerous chemicals are used for textile production, mainly in the wet processing. Of these nearly 2000 different chemicals, many have adverse impacts on health. Some chemicals evaporate, while others are dissolved in treatment water—which ultimately goes back to the environment—and some chemicals remain in the product [31]. Cotton clothing, which, after all, is regarded as particularly natural and healthy, calls for cotton farming, which currently needs 0.2 million tons of pesticides and 8 million tons of fertilizer globally. Although cotton cultivation accounts for only 2.5% of worldwide agricultural land, it is responsible for 16% of global pesticide utilization. Furthermore, the physical health of cotton farmers gets negatively affected from the chemicals used for cultivating cotton, and cases of acute poisoning from pesticides among cotton farmers are commonplace. About 4% of all nitrogen and phosphorus fertilizers used around the world go to cotton production, and these chemicals are a main source of clean water pollution. If merged into the rivers, these chemicals can lead to algal blooms, which starve the river of oxygen. Producing cellulose-based fibers also necessitates large amounts of chemicals and some of the chemicals used are sources of concern. However, the agricultural part of textile production is not solely responsible for the chemical use of the industry. Producing the fibers requires using chemicals too, for example for dyes or finishing treatments. This part of the production is estimated to use approximately 43 million tons of chemicals globally [1].

The microplastics contaminating the oceans are attracting more and more attention from concerned scientists, even though we still do not fully understand its long-term impacts. Microfibers discharged from textiles during washing processes add to the increasing plastic pollution in the oceans [1]. George Leonard, Chief Scientist for The Ocean Conservancy, estimates that the amount of microfibers on seafloor could have reached to the sheer figure of 1.4 million trillion [35].

The waste generated by producing and consuming textiles is another major concern. Textile consumption around the world is calculated to be over 100 million tons [23]. However, the rate of recycling is rather low: Barely 13% of the total material input is in some way recycled after usage. Of this recycled 13%, a minuscule part is used to produce new clothing—less than 1%. The rest is recycled into other, lower-value items such as insulation material, wiping cloths or mattress stuffing [1].

Additionally, odor problems and noise pollution are also negative effects of the textile industry on the environment. Odor pollution is an indicator of environmental change that affects health and human well-being. Odor impacts people by strong, unpleasant or offensive smells that can interfere with one's enjoyment of life especially if they are frequent and/or persistent [36]. When it comes to noise pollution, there are different processes in the textile chain that can produce noise level above 90 dB(A), the allowed limit, and can cause problems especially for the workers. The dry processes produce more noise than the wet processes, due to the fast-moving parts in the processing machines, which is another danger for the workers along with the hearing problems [37].

3. General outlook of the current situation of the textile fibers

The world population has grown tremendously in the past few decades, and the same period also witnessed improvements in living standards in general. These two developments have augmented the consumption of textiles, which in turn increased textile production [38]. The effect of the rising living standards can be seen in the fact that the worldwide consumption of textiles is growing faster than the world population. The demand is expected to grow from around 30 million tons in 1980 to more than 130 million tons in 2025. The figure translates into a growth of over 400%—or an average annual growth rate of 4.3%. In the same period, the world population has been growing by only 1.7% [39].

Global production of all apparel and textile fibers amounted to 110 million tons in 2018, according to the Discover Natural Fibers Initiative (DNFI). This number points to an increase of 4 million tons compared with the previous year and of 35 million tons compared with a decade ago. Natural fibers represent 29%—a 12% decrease since 2008. In 2018, cotton represented 81% of natural fiber production by weight, which overshadows the share of jute, coir and wool, which account for 7%, 3% and 3%, respectively. Cellulosic fiber production represented 6%, synthetic filament 45% and synthetic staple 20% of the total production in 2018. Polyester is the leading synthetic fiber, which represents almost 90% of world filament production and 70% of world synthetic staple production. The rest of the synthetic fibers are mostly composed of nylon, acrylic and polypropylene. However, the figures do not translate into a plunge in the production of natural fibers. The share of natural fibers in total fiber production has decreased in the last decade because the production of polyester has increased exponentially. Synthetic filament production, which is mainly used for the production of fast-fashion apparel, has risen from 26 million tons to 50 million in a decade after 2008, almost doubling in size. During the same period, synthetic staple production increased from 15 million tons to 22 million. Natural fiber production also increased from 2008 to 2018, but the rate was nowhere close to the others: from 31 million tons to 32 million [40].

The global market is prevailed by two types of fibers: polyester, a synthetic fiber, and cotton, a natural fiber. The trend mentioned in the previous paragraph is clearly reflected when the figures for these two fibers are examined. The demand for polyester has doubled—a significant rise that resulted in the fact that polyester has succeeded cotton, the most widely used fiber until the 2000s. Polyester fiber production is estimated to increase to be 3 times more than cotton production in order to meet the still-growing demand, while the production of cotton fibers remained stable. Increasing the production of cotton depends on the land and water resources, which are limited, and the fact that the opportunities to increase yields of cotton cultivation are narrow does not help either. These constraints on cotton production are very significant to understand the growth of the synthetic fiber market [39].

Still, it is important to point out that, despite the growth of synthetic fiber production, cotton, a product with very good fiber characteristics, remains to be considered as the most popular fiber. It is not expected that cotton would largely be replaced or eliminated in the short or medium term from the textile production. Thus, sustainability strategies for cotton will persist to be paramount for the conservation of the environment [25, 41].

Synthetic polymers are mainly produced from petroleum—a nonrenewable resource. But this is not limited to synthetic fibers: Renewable natural polymers such as cotton also depend on nonrenewable resources, because their production needs energy and chemicals that are actually produced from nonrenewable resources. The petroleum reserve of the world might last for several more centuries if the current consumption rates continue, but it does not change the fact that petroleum—like many other natural resources—cannot be replaced in practical terms [42]. Therefore, deciding if natural fibers or manufactured fibers are more eco-friendly is impossible. The production of all types of fibers comes with its own challenges. Some fibers need a lot of water, while others demand lots of energy to produce. The synthetic fibers are not fully biodegradable like the natural and cellulosic fibers, which broadens the waste aspect of the problem. Synthetic fibers are generally petroleum by-products, which makes them nonrenewable materials; however, this gives them the advantage to get conveniently recycled into a good-quality material like polyester—a contrast with cotton, which generally gets down-cycled. But recently, the market has also started to receive recycled, high-quality cotton [19, 20].

Made-By, a nonprofit organization, carried out a study called "The Environmental Benchmark for Fibres." The study focuses on the prevalent fibers in the clothing industry and compares the environmental impact of the production of these materials. The production of the fibers is analyzed from the raw material up until the preparation of the fiber to be spun, thus excluding the later stages, such as spinning itself, fabric manufacturing, dyeing and finishing, garment making, transportation of the product and consumption. The study lists 28 fibers from Class A to Class E (Class A being the most benign) not regarding their quality, durability or performance, but their direct effects on the environment: greenhouse gas emissions, human toxicity, eco-toxicity, energy input, water input and land use [43].

The results demonstrate that all mechanically recycled fibers and organic fibers score "positively," while both natural and synthetic fibers obtained by conventional production methods are far behind in sustainability ranking. Class A materials include mechanically recycled nylon, mechanically recycled polyester, organic flax (linen), organic hemp, recycled cotton and recycled wool, while Class E materials include bamboo viscose, conventional cotton, cuprammonium rayon, generic viscose, rayon, spandex (elastane), virgin nylon and wool [43].

To protect not just human life on Earth, but Earth itself, we have to use the natural resources adequately. The rates at which natural resources are generated and consumed have to be appropriate for the sustainability of the planet. The 2011 annual report of the United Nations Environment Programme (UNEP) predicts the rate of consumption to triple the current rate by the year 2050 [44]. One alternative way to approach the solution of this problem is gradually replacing the traditional linear economy model—which relies on extraction/cultivation of raw materials, use of the product and disposal of the waste in landfills—with the circular material flow—which focuses on reusing and recycling.

4. Textile wastes and recycling

Consumption of textile products has two main aspects that trigger environmental change: the pollution and waste brought about, and the natural resources expended. Pollution is generated not only during the production phase but also during the consumption of the products. The Earth has a natural system that can naturalize pollutants and stabilize a natural equilibrium to a certain extent, but the rate and degree of the release of man-made pollutants into nature challenge this natural equilibrium of the planet [45, 46].

The second factor is the depletion of limited natural resources of the planet through the consumption of goods. Conventional modes of production and consumption dictate utilizing both renewable and nonrenewable resources. Manufacturing processes required during production need natural resources such as fossil fuels (coal, oil and natural gas) to generate energy and raw materials for the actual products (as in the example of plastic, which generally is produced from petrochemicals). Furthermore, in most cases, more natural resources are used up to consume the products themselves. Unfortunately, the utilization of both nonrenewable and renewable resources has a major impact on both localized and global environmental change. As a result of the depletion of resources and generation of pollution, both producing and consuming goods by humans are important sources of environmental change [45, 46].

A significant amount of research has been conducted and published on the environmental impacts of the production and consumption of textiles. The research has helped to inform policy-makers and the public on reducing toxicity of chemicals in production stages, creating industry standards for production and promoting more sustainable ways of cleaning textiles. However, sustainability of the disposal of textiles was not paid much attention until recently.

Textile production is a burden for the environment. Textiles cost significant amounts of natural resources, and the use of toxic chemicals and generation of large quantities of carbon dioxide further augment the problem. However, despite this huge cost, millions of tons of textile products are disposed of every year. In Europe and America, 10 million tons of disposed textile products are predicted to be disposed of, while the estimation for China is double this amount. This textile waste pollutes our environment and clogs landfills around the world on top of all the natural resources used for their production.

Western lifestyle, with its dependence on the culture of consumption, amplifies landfill waste. Not only is the consumption at a high level, but also products are generally overpackaged in the West, which translates into more waste—and to the consumption of natural resources required for packaging. Landfill capacity is not growing at the pace of the increase of the generation of waste, which inevitably means that the cost of waste disposal rises further. This is a major concern for businesses as they need to reduce the overhead costs [38].

The disposal of textile wastes is crucial for the textile industry globally. Tons of textile products get discarded by the consumers and end up in landfills all over the world. Estimates suggest that a vast majority, as high as 95 percent, of the discarded product could in fact be used again—re-worn, reused or recycled, depending on their condition. Indeed, the conditions are pushing the trend, because, as natural resources are limited and cost of waste disposal is increasing, more waste is getting recycled or reused [47].

Textile wastes account for almost 5% of all landfill spaces, according to the US Environmental Protection Agency (USEPA); however, the recycled postconsumer textile wastes are barely 15% annually, and thus, a huge 85% of the waste ends up in landfills. Certain organizations, including the Council for Textile Recycling (CTR), are endeavoring to raise consciousness about keeping the postconsumer textile wastes out of the solid waste streams, with the aim of reaching the level of zero textile waste going to landfills by 2037 [44].

Textile waste is produced through a number of streams including the fiber, textile and clothing manufacturing industry, consumers and the commercial and service industries. CTR categorizes textile recycling material as pre- or postconsumer waste [48, 51].

Preconsumer textile waste, according to CTR, is the waste generated during production—by processing fibers, and the production of finished yarns and textiles, technical textiles, nonwoven, garments and footwear, including offcuts, selvages, shearings, rejected materials and/or B-grade garments. Preconsumer textile waste is usually what is considered as "clean waste." Preconsumer textile wastes are produced by the original manufacturers and never make it to consumers [44, 48].

Postconsumer textile waste refers to textile products that the consumer disposes for any reason—they might be run-down or not liked by the consumer anymore. Generally, postconsumer textile wastes tend to be of good quality, which can be recovered or reused as second-hand clothing, and are generally sold to poorer regions of the world. Even the textile products that will most likely not be used by the consumers can potentially be shredded into fiber to be reused for manufacturing [48].

The fast fashion era has skyrocketed the rate at which textile products are discarded, as "going-out-of-fashion" has become one of the main reasons for "not liking the product anymore." The implementation of a convenient recycling regime can turn these wastes into raw materials to be used in producing future, value-added products. This is the current aim for the ongoing development of textile waste management systems, which seek to produce value-added products through recycling [44]. Textile waste treatment strategies include reducing, reusing and recycling, as shown in **Figure 1**. The first and most preferred approach, reducing, is aimed at, if possible, avoiding any waste entirely. The second approach, reusing, aims literally for the item to be reused by a consumer after it has been discarded by another. The third and last approach is recycling: the materials of discarded items are transformed into new products [49]. Wastes can be recycled to products for the same purpose with their first use, or they can be upcycled or downcycled. In upcycling, wastes are converted into high-value products with different purposes than the original use, while in downcycling, valuable products are converted into lower-value materials [50].

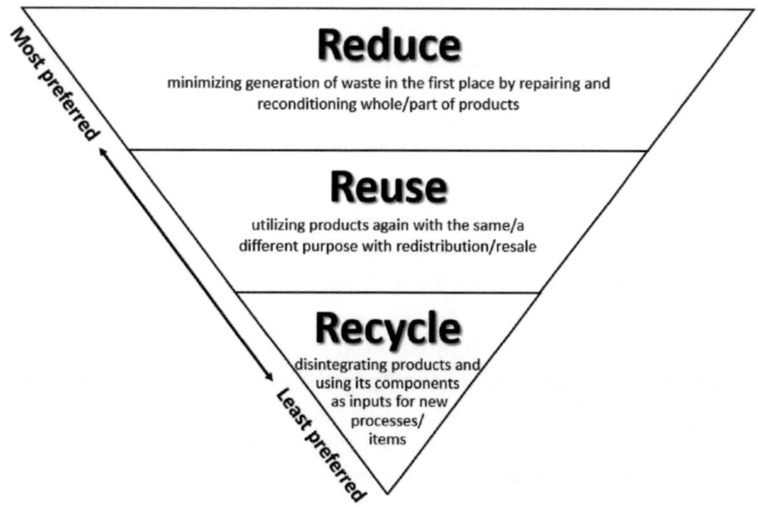

Figure 1.
Textile waste treatment strategies 3R concept.

Sustainability aspires to derive maximum benefit from products by extending their life. Studies conducted by economists and environmentalists on technical and economic requirements for sustainability reveal that it is imperative to reduce waste generation and increase recycling. Below are a few reasons why recycling is important [51, 52]:

- Economical reasons: recycling programs cost less than waste disposal programs. The high water, energy and manufacturing consumption makes it much cheaper to recycle than to produce some new textile products. Recycling can be made financially rewarding, as people can receive money for turning in certain recyclable products.

- Social reasons: recycling creates jobs. Recycling centers create four jobs for every one job in the waste disposal industry. The method can also create opportunities for small businesses.

- Environmental reasons: recycling conserves natural resources such as water, oil and natural gas; saves energy as it requires less energy compared with manufacturing brand new products; produces less greenhouse gases; and prevents the destruction of natural habitats.

Recycling and recovery of textile products are not as common as the material groups such as glass, metal, plastic and paper and product groups such as electronic, packaging and automotive. Recycling activities related to the textile sector are mainly focused on the treatment of chemical wastes and polluted water—problems that arise during production processes. The number of scientific or practical studies on the recycling of solid wastes is extremely limited [53].

Even though the textile and apparel sector is one of the most intense consumption sectors, implementation of recycling throughout the sector is not satisfying. However, parallel to the increasing global awareness of environmental problems, the awareness of consumers about sustainability has also started to increase. Consumers are now demanding recycled textile products and manufacturers are seeking ways to meet this demand [53, 54].

The waste generated by the textile sector contributes to land, water and air pollution. Decomposing textiles generate greenhouse gases and thus air pollution. The vast amount of chemicals used for producing textile goods unavoidably pollutes the rivers. And discarded textile products fill up landfills, which are already scarce. All these wastes are resources that could have been used to create value-added products. Not only this potential is lost, but also more raw materials are required to be used, which in turn results in more energy to be consumed [55].

Wasted materials can be recovered through reusing a product as is and converting the waste into a product. A material should get to be reused as much as possible and the consumer finally decides to discard it, and then recycling would be a good alternative to reduce the carbon footprint [56].

Recycling technologies tend to be divided into primary, secondary, tertiary and quaternary approaches, and all these four methods are applicable to recycle fibers. Primary approaches refer to the process of recycling a material to what it was originally. Secondary recycling means melt processing a plastic product into a lower-quality but nevertheless a new one. Tertiary recycling refers to processes that convert the plastic wastes into basic chemicals or fuel, such as pyrolysis and hydrolysis. Quaternary recycling involves burning the fibrous solid waste and converting it into a source of energy exploiting the heat generated through burning [38, 57].

Obviously, the most fruitful method of recycling is the primary approach. This approach, also called closed-loop recycling, is only applicable to man-made fibers such as PET or PA. This primary approach includes collecting textile waste discarded by the user and using this in new clothing as material for the production of yarn. The most common method of recycling actually is what is called open-loop recycling. In this method, the output material does not have a high-enough level of quality to produce new clothes; thus, it gets downgraded. The study on cotton fibers by Ütebay et al. demonstrates the deterioration in fiber quality. Downgraded material can be used as mattress upholstery or isolation material in cars. Through open-loop recycling, some value is recuperated from the textile waste, which would otherwise have been incinerated. However, this does not help to reduce the necessity of raw materials to produce clothing. Therefore, closed-loop recycling remains to be an attractive alternative. A closed-loop supply chain provides the advantage to recover more value from used products [56, 58, 59].

The most recycled textile waste is thermoplastic polymer-based fibers because they are easy to process and can be given different forms and shapes afterwards. Nevertheless, natural fibers such as cotton, wool and silk are also finding their ways into the recycling stream through downcycling or upcycling [44].

Recycling of textiles today is not a wide industry. The number of companies that offer services of recycling of textile fibers is limited because it is economically not beneficial and technologically not advanced. The lack of technological innovation and the continuing supply of cheap fabrics into the markets hinder the motivation for research, development and application of recycling techniques; however, it does not totally block the development of new technologies for recycling textile waste. Soon, certain obstacles will need to be faced and resolved to further increase textile waste recycling [10, 60].

In terms of technology, the fundamental question is the fiber composition of textile materials. The current garments in today's markets vary more in terms of design and fiber content than in the past. The other compounded factors are issues/difficulties in separating the blended components, efficiency of separation, quality of separated material and hence the recycled material's quality and so on. As recycled fibers and fabrics still provide a low level of quality, virgin natural and synthetic fibers remain to be popular options [10, 38, 56].

There are some recycling technologies available on the market today. Others are on the way, albeit few. Other changes need to accompany these research and trial endeavors—in terms of economy, processing costs should be reduced; in terms of policy, relevant standards should be implemented; and in terms of governance, textile waste should be recollected much more efficiently. Apart from increasing the efficiency of recycling methods and processes, the market for recycled products should grow. In short, recycling needs more encouragement, wherever it is economically and technically feasible [47].

There are important benefits of recycling textiles, both environmental and economical. It reduces the need for landfill space, consumption of already scarce virgin resources, pollution as well as water and energy consumption and the demand for dyes and fixing agents [61]. However, even though recycling offers ways to reduce environmental negative impacts, it is not exempt from certain problems. Wang [38] lists the following as challenges:

- The mechanical, chemical or biological processes to recycle waste still require a certain amount of energy.

- The recycling processes continue to require new raw material input.

- The recycling processes still generate emissions into air, water and land.

Evidently, recycling is not always the preferred approach, when not only the environmental context but also the competitiveness of the final product in the market is taken into account. The existing recycling technologies need to get better, cleaner, more energy efficient and less costly [38].

Textile recycling, a key concept for sustainability, currently faces hurdles related to cost, time and technology. But as sustainability becomes more and more important, more initiatives are getting incentivized and sponsored by both manufacturers and other organizations in the textile sector to help advance the results of textile recycling. This tendency can be seen through the fact that certain textile recycling companies have shown promising growth. Different strategies and policies were coined in different regions to promote an efficient way of recycling for conserving the environment more and increasing the economic efficiency [38].

Many voluntary and nonprofit organizations run campaigns to conserve natural resources by creating awareness of both downcycling and upcycling recycling

Figure 2.
Labels for recycled products.

concepts. The campaigns aim to convince the consumers that using recycled products is an esteemed way of adding value to oneself, the product and the world.

Examples of commercially available labels for recycled products are given in **Figure 2** [62–68]. All these initiatives are expected to promote environmentally appropriate, socially beneficial and economically viable management systems [44].

In addition to these certifications, some standards such as Social Responsibility Standard (ISO 26000), Environmental Management System (ISO 14001), Occupational Health and Safety Management System (ISO 45001) and Energy Management System (ISO 50001) also support sustainability and social responsibility practices and contribute to their dissemination [69–72].

An unsustainable consumption of textile goods ensures the deterioration of the environmental degradation. To achieve environmental integrity and sustainability, incentivizing the textile companies to produce more environment-friendly products is not enough—the behavior of consumers also needs to change, creating more awareness toward the conservation of the environment [45]. In this context, Connolly and Prothero [73] write: "Rather than focus on the issue of whether green consumption can work as a strategy, we should perhaps try to gain a greater understanding of the process that has led people to believe that they, as individuals, can help solve global environmental problems."

5. Conclusion

The fibers obtained by recycling are generally evaluated in the production of lower-value products (downcycling) compared to the original product. However, nowadays, recycling fibers have started to increase their evaluation in high value-added products (downcycling). On the other hand, the perspective that focuses only on the cost aspect of the production of recycled garments is not correct. Considering water consumption as well as pesticides and artificial fertilizers used, the textile industry is known to be one of the most polluting and waste-generating sectors. From this point of view, recycling of textiles and garments is of great importance in terms of reducing the use of natural resources (e.g., water used to grow seeds or oil used in the production of synthetic fibers) and CO_2 emissions. Recycling will also save energy and chemicals to produce new textiles, as well as prevent pollution from the production process. In this context, it is important for the future of our world to review all production and consumption processes and supply chains in the focus of circular economy and sustainability. Therefore, the recycling of textile industry waste is very important.

The future of textile recycling mostly depends on its implementation in the industry and gaining more experience and grounds for more innovative methods. Clothing retailers are key actors on this front, as they are uniquely positioned to be able to influence and improve consumers' approach in favor of sustainability. Not only do clothing retailers have the potential to influence consumer decisions, but also they are in a position to alter consumption patterns. People can learn the importance of recycling as well as reuse and resales by the help of companies, and this is not limited to developing countries. Through such actions, consumer awareness about sustainable consumption would increase, leading to less environmental damage in the future.

Author details

Burçin Ütebay[1]*, Pinar Çelik[2] and Ahmet Çay[2]

1 Üniteks R&D Center, İzmir, Turkey

2 Faculty of Engineering, Department of Textile Engineering, Ege University, İzmir, Turkey

*Address all correspondence to: burcin@uniteks.com.tr

References

[1] Ellen MacArthur Foundation. A New Textiles Economy: Redesigning Fashion's Future [Internet]. 2017. Available from: https://www.ellenmacarthurfoundation.org/assets/downloads/publications/A-New-Textiles-Economy_Full-Report.pdf [Accessed: 28 January 2020]

[2] Guldmann E. Best practice examples of circular business models. Denmark: The Danish Environmental Protection Agency; 2016. p. 56. ISBN: 978-87-93435-86

[3] Ellen MacArthur Foundation. What Is the Circular Economy? [Internet]. Available from: https://www.ellenmacarthurfoundation.org/CIRCULAR-ECONOMY/WHAT-IS-THE-CIRCULAR-ECONOMY [Accessed: 28 January 2020]

[4] Ekström KM, Salomonson N. Reuse and recycling of clothing and textiles—A network approach. Journal of Macromarketing. 2014;**34**:1-17. DOI: 10.1177/0276146714529658

[5] Crane D. Introduction. In: Gonza'lez AM, Bovone L, editors. Fashion and Identity: A Multidisciplinary Approach. New York: Social Trends Institute; 2007. pp. 15-20. ISBN: 9780857850577

[6] AM G'l. Fashion, image, identity. In: AM G'l, Bovone L, editors. Fashion and Identity: A Multidisciplinary Approach. New York: Social Trends Institute; 2007. pp. 39-66. ISBN: 9780857850577

[7] Kawamura Y. Adoption and consumption of fashion. In: Kawamura Y, editor. Fashion-Ology: An Introduction to Fashion Studies. Berg: Oxford; 2005. pp. 89-105. DOI: 10.2752/9781847888730

[8] Östberg J, Kaijser L. Konsumtion. Malmö: Liber AB; 2010. p. 128. ISBN: 9789147088744

[9] Slater D. Consumer Culture and Modernity. Cambridge: Polity Press; 1997. p. 240. ISBN: 9780745603049

[10] Fletcher K. Sustainable Fashion and Textiles: Design Journeys. London: Earthscan; 2008. p. 239. ISBN: 9781844074815

[11] Pasquinelli I. What Can the Fashion Industry Do to Become more Sustainable? [Internet]. 2012. Available from: https://www.theguardian.com/sustainable-business/fashion-industry-sustainability-strategy [Accessed: 28 January 2020]

[12] Welters L, Lillethun A. The Fashion Reader. New York: Berg Publishers; 2007. p. 480. ISBN: 9781845204860

[13] Bruce M, Daly L. Buyer behaviour for fast fashion. Journal of Fashion Marketing and Management. 2006;**10**(3):329-344. DOI: 10.1108/13612020610679303

[14] Johansson E. Slow fashion—The answer for a sustainable fashion industry? [thesis]. Borås: The Swedish School of Textiles; 2010

[15] Shen B, Wang Y, Lo C, Shum M. The impact of ethical fashion on consumer purchase behavior. Journal of Fashion Marketing and Management. 2012;**16**(2):234-245. DOI: 10.1108/13612021211222842

[16] Lo CKY, Yeung ACL, Cheng TCE. The impact of environmental management systems on financial performance in fashion and textiles industries. International Journal of Production Economics. 2012;**135**(2):561-567. DOI: 10.1016/j.ijpe.2011.05.010

[17] Chan TY, Wong CWY. The consumption side of sustainable fashion supply chain: Understanding fashion consumer eco-fashion consumption

decision. Journal of Fashion Marketing and Management. 2012;**16**(2):193-215. DOI: 10.1108/13612021211222824

[18] Thomas S. The "green blur" to ecofashion: Fashioning an eco-lexicon. Fashion Theory. 2008;**12**(4):525-540. DOI: 10.2752/175174108X346977

[19] Niinimäki K. Tenents of sustainable fashion. In: Niinimäki K, editor. Sustainable Fashion: New Approaches. Helsinki: Aalto ARTS Books; 2013. pp. 12-31. ISBN: 9789526055725

[20] Chen H, Burns LD. Environmental analysis of textile products. Clothing and Textile Research Journal. 2006;**24**(3):248-261. DOI: 10.1177/0887302X06293065

[21] Zion Market Research. Textile Market by Raw Material (Cotton, Wool, Silk, and Chemicals & Synthetic), by Product Type (Natural Fiber, Polyester, and Nylon), and by Application (Household, Technical & Workwear, Fashion & Clothing, and Others): Global Industry Perspective, Comprehensive Analysis, and Forecast, 2018-2025 [Internet]. 2019. Available from: https://www.zionmarketresearch.com/market-analysis/textile-market [Accessed: 28 January 2020]

[22] Solidarity Center. Global Garment and Textile Industries: Workers, Rights and Working Conditions [Internet]. 2019. Available from: https://www.solidaritycenter.org/wp-content/uploads/2019/08/Garment-Textile-Industry-Fact-Sheet.8.2019.pdf [Accessed: 28 January 2020]

[23] Harder J. Throwaway Clothing: The Limitations of Textile Recycling [Internet]. 2019. Available from: https://www.recovery-worldwide.com/download/1473851/recovery_4_2019_online.pdf [Accessed: 28 January 2020]

[24] Goldbach M, Seuring SA, Back S. Co-ordinating sustainable cotton chains for the massmarket: The case of the german mail-order business OTTO. Greener Management International. 2003;**43**:65-78. DOI: 10.9774/GLEAF.3062.2003.au.00008

[25] Hansen EG, Schaltegger S. 100 per cent organic? A sustainable entrepreneurship perspective on the diffusion of organic clothing. Corporate Governance. 2013;**13**:583-598. DOI: 10.1108/CG-06-2013-0074

[26] Hansen EG, Große-Dunker F, Reichwald R. Sustainability innovation cube—A framework to evaluate sustainability-oriented innovations. International Journal of Innovation Management. 2009;**13**(4):683-713. DOI: 10.1142/S1363919609002479

[27] United Nations. Report of the World Commission on Environment and Development: Our Common Future [Internet]. 1987. Available from: http://www.un-documents.net/wced-ocf.htm [Accessed: 28 January 2020]

[28] Fletcher K. Systems change for sustainability in textiles. In: Blackburn RS, editor. Sustainable Textiles: Life Cycle and Environmental Impact. Cambridge: Woodhead; 2009. pp. 369-380. ISBN:9781845694531

[29] Bhuiyan MR, Rahman MM, Shaid A, Bashar M, Khan MA. Scope of reusing and recycling the textile wastewater after treatment with gamma radiation. Journal of Cleaner Production. 2016;**112**:3063-3071. DOI: 10.1016/j.jclepro.2015.10.029

[30] Raja ASM, Arputharaj A, Saxena S, Patil PG. Water requirement and sustainability of textile processing industries. In: Muthu SS, editor. Water in Textiles and Fashion: Consumption, Footprint, and Life Cycle Assessment. Cambridge: Woodhead; 2019. pp. 155-173. DOI: 10.1016/B978-0-08-102633-5.00009-9

[31] Choudhury AKR. Environmental impacts of the textile industry and its assessment through life cycle assessment. In: Muthu SS, editor. Roadmap to Sustainable Textiles and Clothing—Environmental and Social Aspects of Textiles and Clothing Supply Chain. Singapore: Springer Science-Business Media; 2014. pp. 1-39. DOI: 10.1007/978-981-287-110-7_1

[32] Nagaraj AR. Energy Management in Textile Industry [Internet]. 2012. Available from: https://www.fibre2fashion.com/industry-article/6276/energy-management [Accessed: 28 January 2020]

[33] Reddy BS, Ray BK. Understanding industrial energy use: Physical energy intensity changes in Indian manufacturing sector. Energy Policy. 2011;**39**(11):7234-7243. DOI: 10.1016/j.enpol.2011.08.044

[34] Greenpeace International, Dirty Laundry: Unravelling the Corporate Connections to Toxic Water Pollution in China [Internet]. 2011. Available from: http://p3-raw.greenpeace.org/international/en/publications/reports/Dirty-Laundry/ [Accessed: 28 January 2020]

[35] Leonard GH. Oceans, Microfibers and the Outdoor Industry: A Leadership Opportunity [Internet]. 2016. Available from: http://storyofstuff.org/wp-content/uploads/2017/01/Oceans-Microfibers-and-the-Outdoor-Industry.pdf [Accessed: 28 January 2020]

[36] Pooja VP, Pooja TL, Anil CR. Odour pollution and its measurement. International Journal for Research in Applied Science & Engineering Technology (IJRASET). 2015;**3**(12):221-229. ISSN: 2321-9653

[37] Farooq A. Vulnerability Risk Assessment to Gauge Impact of Inefficient and Unsustainable Practices in the Leather and Textile Sectors [Internet]. 2018. Available from: https://d2ouvy59p0dg6k.cloudfront.net/downloads/vulnerabilityriskassessment_w.pdf [Accessed: 16 March 2020]

[38] Wang Y, editor. Recycling in Textiles. 1st ed. Cambridge: Woodhead Publishing; 2006. p. 248. ISBN: 9781855739529

[39] Qin Y. Global Fibres Overview [Internet]. 2014. Available from: https://www.orbichem.com/userfiles/APIC%202014/APIC2014_Yang_Qin.pdf [Accessed: 28 January 2020]

[40] Townsend T. Natural Fibres and the World Economy July 2019 [Internet]. 2019. Available from: https://dnfi.org/coir/natural-fibres-and-the-world-economy-july-2019_18043/ [Accessed: 28 January 2020]

[41] Die Umweltberatung. Handbuch Ecofashion—Mode mit Zukunft! [Internet]. 2009. Available from: https://www.umweltberatung.at/download/?id=ecofashion_handbuch-3051-umweltberatung.pdf [Accessed: 28 January 2020]

[42] Wang Y. Fiber and textile waste utilization. Waste Biomass Valor. 2010;**1**(1):135-143. DOI: 10.1007/s12649-009-9005-y

[43] Made-By. Environmental Benchmark for Fibres [Internet]. 2013. Available from: https://www.commonobjective.co/article/made-by-environmental-benchmark-for-fibres [Accessed: 28 January 2020]

[44] Vadicherla T, Saravanan D. Textiles and apparel development using recycled and reclaimed fibers. In: Muthu SS, editor. Roadmap to Sustainable Textiles and Clothing Eco-Friendly Raw Materials, Technologies, and Processing Methods. Singapore: Springer Science-Business Media; 2014. pp. 139-160. DOI: 10.1007/978-981-287-065-0_5

[45] Connell KYH, Kozar JM. Environmentally sustainable clothing consumption: Knowledge, attitudes, and behavior. In: Muthu SS, editor. Roadmap to Sustainable Textiles and Clothing—Environmental and Social Aspects of Textiles and Clothing Supply Chain. Singapore: Springer Science-Business Media; 2014. pp. 41-61. DOI: 10.1007/978-981-287-110-7_2

[46] Chiras DD, editor. Environmental Science: A Systems Approach to Sustainable Development. 5th ed. Belmont: Wadsworth; 1998. p. 608. ISBN: 0534538746 9780534538743

[47] Lu JJ, Hamouda H. Current status of fiber waste recycling and its future. Advanced Materials Research. 2014;**878**:122-131. DOI: 10.4028/www.scientific.net/AMR.878.122

[48] Sakthivel S, Ramachandran T, Gowthami R, Chandhanu R, Padma Priya J, Vadivel P, Gokulraj S. Source and Effective Utilization of Textile Waste in Tirupur [Internet]. 2010. Available from: https://www.fibre2fashion.com/industry-article/5251/source-and-effective-utilization-of-textile-waste-in-tirupur [Accessed: 28 January 2020]

[49] Petzet M. Reduce Reuse Recycle [Internet]. 2012. Available from: http://www.reduce-reuse-recycle.de/pdf/en/RRR_Bienn2012_Paper_August2012.pdf [Accessed: 16 March 2020]

[50] The difference between recycling, upcycling and downcycling [Internet]. 2012. Available from: https://theupcyclingfashionista.wordpress.com/2013/11/08/the-difference-between-recycling-upcycling-and-downcycling/ [Accessed: 16 March 2020]

[51] Cuc S, Vidovic M. Environmental sustainability through clothing recycling. Operations And Supply Chain Management. 2011;**4**(2):108-115. DOI: 10.31387/oscm0100064

[52] Palm D. Improved Waste Management of Textiles [Internet]. 2011. Available from: http://www.hallbaravfallshantering.se/download/18.5c577972135ee95b56380001388/1350483390437/Palm+(2011)+Improved+waste+management+of+textiles+(B1976).pdf [Accessed: 28 January 2020]

[53] Güngör A, Palamutçu S, İkiz Y. Cotton textiles and the environment: Life cycle assessment of a bathrobe. Tekstil ve Konfeksiyon. 2009;**19**(3):197-205

[54] Kurtoğlu Necef Ö, Seventekin N, Pamuk M. A study on recycling the fabric scraps in apparel manufacturing industry. Tekstil ve Konfeksiyon. 2013;**23**(3):286-289

[55] Woolard R. Logistical model for closed loop recycling of textile materials [thesis]. North Carolina: Textile Engineering; 2009

[56] Muthu SSK, Li Y, Hu JY, Ze L. Carbon footprint reduction in the textile process chain: Recycling of textile materials. Fibers and Polymers. 2012;**13**(8):1065-1070. DOI: 10.1007/s12221-012-1065-0

[57] Wang Y, Wu HC, Li VC. Concrete reinforcement with recycled fibers. Journal of Materials in Civil Engineering. 2000;**12**(4):314-319. DOI: 10.1061/(ASCE)0899-1561(2000)12:4(314)

[58] Ütebay B, Çelik P, Çay A. Effects of cotton textile waste properties on recycled fibre quality. Journal of Cleaner Production. 2019;**222**:29-35. DOI: 0.1016/j.jclepro.2019.03.033

[59] Hagoort S. Evaluating the impact of closed loop supply chains on Nike's environmental performance and costs [thesis]. Eindhoven: Operations Management and Logistics at Eindhoven University of Technology; 2013

[60] Muthu SS, Li Y, Hu JY, Mok PY. Recyclability potential index (RPI): The concept and quantification of RPI for textile fibres. Ecological Indicators. 2012;**18**:58-62. DOI: 10.1016/j.ecolind.2011.10.003

[61] Eryuruk SH. Greening of the textile and clothing industry. Fibres & Textiles in Eastern Europe. 2012;**95**(6):22-27

[62] Textile Exchange. Global Recycled Standard 4.0 [Internet]. 2017. Available from: https://textileexchange.org/wp-content/uploads/2017/06/Global-Recycled-Standard-v4.0.pdf [Accessed: 28 January 2020]

[63] Textile Exchange. Recycled Claim Standard 2.0 [Internet]. 2017. Available from: https://textileexchange.org/wp-content/uploads/2017/06/Recycled-Claim-Standard-v2.0.pdf [Accessed: 28 January 2020]

[64] Textile Standards & Legislation [Internet]. Available from: www.textilestandards.com [Accessed: 28 January 2020]

[65] FSC Labels [Internet]. Available from: https://fsc.org/en/page/fsc-labels [Accessed: 28 January 2020]

[66] Federal Trade Commission. Consumer Perception of "Recycled Content" and "Organic" Claims [Internet]. 2016. Available from: https://www.ftc.gov/system/files/documents/reports/consumer-perception-recycled-content-organic-claims-joint-staff-report-federal-trade-commission/consumer_perception_of_recycled_content_and_organic_2016-08-10.pdf [Accessed: 28 January 2020]

[67] Redress. What Is the R Cert? [Internet]. Available from: https://www.redress.com.hk/r-cert/about [Accessed: 28 January 2020]

[68] SCS Global Services. Recycled Content Certification [Internet]. Available from: https://www.scsglobalservices.com/services/recycled-content-certification [Accessed: 28 January 2020]

[69] Jasıulewıcz-Kaczmarek M. Sustainable development and new revision of ISO management standards. In: Proceedings of the 3rd International Conference on Social Science (ICOSS 2016); 19-20 September 2016; Bali. DOI: 10.12783/dtssehs/icss2016/9113

[70] Borella IL, Carvalho Borella MR. Environmental impact and sustainable development: An analysis in the context of standards ISO 9001, ISO 14001, and OHSAS 18001. Environmental Quality Management. 2016;**25**(3):67-83. DOI: 10.1002/tqem.21460

[71] Medina-Serrano R, González-Ramírez M, Gascó-Gascó JL, Llopis-Taverner J. Sustainable supplier evaluation practices across the supply chain. Dirección y Organización. 2019;**69**:13-26. ISSN: 2171-6323

[72] Wiengarten F, Humphreys P, Onofrei G, Fynes B. The adoption of multiple certification standards: Perceived performance implications of quality, environmental and health & safety certifications. Production Planning & Control. 2017;**28**(2):131-141. DOI: 10.1080/09537287.2016.1239847

[73] Connolly J, Prothero A. Green consumption: Life-politics, risk and contradictions. Journal of Consumer Culture. 2008;**8**(1):117-145. DOI: 10.1177/1469540507086422

Chapter 4

Understanding Denim Recycling: A Quantitative Study with Lifecycle Assessment Methodology

Sedef Uncu Akı, Cevza Candan, Banu Nergis and Neslihan Sebla Önder

Abstract

Global denim jeans market size is over 70 billion dollars today, and it continues to grow with a compound annual growth rate (CAGR) of 2%. As a reflection of this annual growth, systems' waste generation also increases. Textile waste including denim accounts for nearly 5% of all landfill space, according to the Environmental Protection Agency (EPA). Today, efficient recycling technologies are essential to revive the waste generated. Recycling technologies potentially represent a new way to engineer products. This book chapter analyzes these different recycling technologies and their advantages and challenges and concentrates on denim fabrics produced with recycled content. Life cycle assessment (LCA) methodology is adopted to quantify and compare the environmental impact of recycled denim fabrics. The chapter concludes the challenges and the future of recycling, creating systems, and engineering of waste. The view of seeing waste as a raw material potentially represents a new opportunity to design circular systems.

Keywords: denim recycling, textile waste, life cycle assessment, design for sustainability, circularity

1. Introduction

Denim is one of the most prevalent fabrics worldwide, and therefore the environmental degradation associated with denim jean production is highly dangerous as the volume of jeans produced and used by consumers today is humongous. Also, as cotton is the fundamental building block for denim production, denim represents a mainspring for cotton consumption. Considering the fact that the textile and apparel industry in general is pressured to increase the recycling potential for cotton to cover contemporary and prospective market demands, the use of recycled materials (fiber, yarn, and fabric) in the production of new denim products has become an area of great interest. The decision to use recycled materials in products occurs during design and product development, and it continues throughout the manufacturing processes. The most important requirements for recyclable denim designs are monomaterials, elimination of toxic substances, modular manufacturing for ease of removal or exchange, easily compatible materials, and labels or codes that can be recycled.

The environmental impact of denim products, during manufacture, use, or disposal, can be evaluated by looking at the different phases of the product's life cycle and taking action at the phases where it will be most effective to reduce the impact. However, the lifecycle of a product is long and complicated, covering many areas with many people involved in each phase [1]. There is a variety of voluntary and mandatory tools which help achieve this objective. These include economic instruments, bans on certain substances, environmental labeling, voluntary agreements, product design guidelines, etc. Life cycle assessment (LCA) methodology is one of the aforementioned tools, and within the frame of this chapter, it is adopted to quantify and compare the environmental impact of recycled denim fabrics. But before that, the global denim market, environmental impacts of denim manufacturing, sources of denim waste, and recycling processes currently used in textiles and apparel today are also discussed in this chapter.

2. An overview of the global denim market

Denim is a type of sturdy cotton twill woven fabric with a characteristically diagonal ribbing known for its use in the form of jeans and other clothing all across the world. Originally used for workers' clothes, denim has entered the world of high fashion for a long time, and this has urged the industry to produce innovative denim with different fibers including lyocell, bamboo, viscose, rayon, polyethylene terephthalate, and elastane. Also, several types of washes have been introduced, such as stonewash and stoneless wash, acid wash, moon wash, monkey wash, and enzymes which have been used because of their environmental-friendly nature in comparison with hypochlorite and pumice stones [2].

Rising casualization of clothing all over the world, together with the availability of high-quality, cost-effective denim products, has contributed to the growth of the market over the years. E-commerce is the other factor which has a positive influence on the denim market's growth. In addition, brands have started to reinforce their commitments to responsible production efforts and launched more sustainable collections to the market, which has dramatically increased new retail products described as "sustainable" since 2017. Orta Anadolu, for example, introduced its organic cotton production of denim fabrics in 2006 and a capsule collection without virgin cotton in 2018. Similarly, Diesel launched a sustainable capsule collection with Coca-Cola [3].

The global denim market is segmented into North America (NA), Europe (EU), Asia-Pacific (APAC), the Middle East and Africa (MEA), and South America (SA). The major companies operating in the global denim market are Levi Strauss & Co., VF Corporation (with renowned brand names of Lee and Wrangler), Diesel SpA, Gap Inc., Hennes & Mauritz AB, Inditex, and PVH Corp., and Tommy Hilfiger Licensing LLC. The jean market is a highly fragmented market, with a strong presence of local and global players operating all over the world. Thus, to sustain their positions in the market, the active players are bringing innovation in their product offering, in order to cater to changing consumers' fashion lifestyle [4].

Also, they have been focusing on social media platforms as well as online distribution channels for their marketing and branding activities so as to attract more customers. Denim manufacturers, however, appear to focus on acquisitions, expansions, and collaborations with mostly startup companies to gain a significant market share [2–3].

In 2019, the global denim market was valued at approximately 90 billion US dollars and is expected to reach a value of around 107 billion US dollars by 2023 [5]. The global denim jean market size, on the other hand, was valued at USD 64.62 billion in 2018, and it is estimated to expand further at a compound annual growth rate (CAGR) of 6.81% from 2019 to 2025 [6]. The Asia-Pacific region led the market in 2019, which was followed by North America, Europe, South America, and the Middle East as a response to the introduction of global brands and an emphasis on premium denim. Japan, China and India are expected to lead the region through 2024 as a result of the countries' rise in promotional activities within the denim market [2, 4].

According to the World Denim Fabric Foreign Trade Report published by the Istanbul Textile and Raw Materials Exporters Association (ITHIB), global denim fabric exports increased by 5.1% in 2018 and reached approximately 5 billion US dollars. The top five countries performing most denim exports are China (42% share of total exports), Pakistan (11.7% share of total exports), India (8.2% share of total exports), Hong Kong (7.4% share of total exports), and Turkey (6.8% share of total exports). It is stated that there is a strong growth in the denim production in Bangladesh and Vietnam, although these countries are mostly importing denim fabric [7].

Europe, the United States, and Japan are the biggest consumers of denim, whereas China and India, being relatively younger economies, are witnessing a steady rise in demand for denim. In 2018, men's wear segment accounted for the largest market share of more than 55%. It is forecasted that the segment will retain its leading position over the upcoming years mainly because of the improved standards of living and demand for trendy fashion apparels. Women's wear segment is, on the other hand, expected to register the fastest compound annual growth rate (CAGR) of 7.25% from 2019 to 2025. This growth is attributed to the high product demand, especially in emerging countries. In addition to that, constant product innovation in this segment is also expected to drive the growth further [2, 6].

2.1 Environmental impacts of denim processing

It is estimated that producing a pair of jeans consumes around 2900 liters of water and large amounts of chemicals and energy. If this is multiplied by the number of jeans produced globally, one can get an idea of the enormous contribution of wastewater and harmful gases by denim industry to the environment [8, 9].

Indigo dye is one of the organic colorants used to color textiles, paper, leather, and plastic and for many applications such as cosmetic and photochemical production. Unfortunately, textile effluents containing indigo dye and other dye types make water toxic and harmful for human and animal consumption, which causes an imbalance in different aquatic ecosystem food chains [10]. The use of synthetic indigo and sulfur dyes also presents serious effluent problem. Bearing in mind that majority of warp dyeing for denim uses indigo and sulfur dyes, the environmental impacts of denim processing can be classified into three main categories:

- Water pollution: dyeing and finishing effluent discharge in water bodies
- Air pollution: cotton dust, abrasives, and chemicals found in air
- Solid waste (sludge)

As is well-known, denim washing is imparted to fabric to improve the softness, comfort, and most importantly achieving a variety of looks such as a faded or worn-out appearance. Pumice stone is used to stonewash denim garments. The stone gets abraded during the process and becomes powdered; part of it remains in the liquor, and part of it sticks to the garment. A sizeable amount of water is required for repeated washing cycles to remove the deposited pumice from the denim. The effluent and pumice dust lead to environmental pollution. Sandblasting is a mechanical finish which uses sand containing silica. The minute silica dust spreads in air, it poses serious respiratory disease such as silicosis [10].

Micro sanding is another finish which pollutes the environment. In the case of chemical washing involving the use of sodium hypochlorite or potassium permanganate, the effluent contains chlorinated organic substances which cause severe impacts to the environment, and the bleaching chemicals are harmful to human health. Acid wash uses both pumice stone and chemicals, namely, sodium hypochlorite or potassium permanganate, and it does not require water but leads to pollution through the effluent having pumice dust and residual manganese which are hazardous [11–12].

Despite all these setbacks of the denim processes summarized above, various suitable treatment processes have been developed and employed for the dyeing effluent on the basis of the nature and complexity of the dyes and chemicals present in denim. Fortunately, many denim companies and their suppliers have been striving hard to embrace greener methods such as laser processing and nano bubble ozone washing machines and are also making effort to develop new techniques of producing jeans, as a part of their business strategies to preserve the environment. They have also understood the importance and the need to build a sustainable business [12–14].

2.2 Sources of denim waste

It is considered that sustainable material management is a precursor of circular economy, which promotes recycling, reuse, and remanufacturing. It was estimated that around 65 billion tons of raw materials were processed by the industrial system at the end of the first decade of the twenty-first century, and this quantity is expected to reach about 82 billion tons by the end of 2020. Therefore, in the last two decades, circular economy (CE) is gaining growing global consideration as the new development model which is capable of influencing the existing production and consumption model [15]. Within that concept, waste is classified on the basis of generation as pre-consumer textile waste, post-consumer textile waste, and industrial textile waste. Pre-consumer waste is the remaining production processes in the industry which includes raw material to finished products ready for market. This may include offcuts, shearing, selvedges, b-grade garments, export surplus, etc. which are homogenous and clean in nature to be used for other purposes. The waste under this class has great potential for reuse and recycling. The post-consumer textile waste can include any product that has completed its life cycle and is no longer useful to the consumer in both function and esthetics. Industrial textile waste is, however, the result of the manufacturing processes and is termed as dirty waste [13, 16].

The textile and apparel industry, which generates a substantial environmental footprint from cultivation, fabric and garment manufacturing, to the landfill disposal of post-consumer items, faces tremendous environmental and resource challenges [17]. In order to tackle such challenges to some extent, several fashion companies offer their customers to take care of their worn-out clothes including denim jeans. However, studies have concluded that less than 1% of these collected clothes are being recycled while nearly 80% of them are mainly sold on the second-hand market in poor countries around the world or used as blankets or isolation material. The remaining 20% of the clothes are either sold on the second-hand market within the EU or is sent to landfill or incineration [18].

Being the largest fraction of apparel, waste jeans (or waste denim fabric from tailoring operations) are composed mainly of "cotton" and "cotton/polyester" fabric with different weight ratios [19]. Although this may lead to the understanding that denim made with 100% cotton, in particular, will readily deteriorate in the environment, in practice a pair of such jeans can stay alive in the environment for a very long time, and therefore the negative environmental impact is very high. The literature reveals that the amount of waste jeans generated annually is estimated as 2.16 million tons and only 35–50% of this amount is collected in Western Europe in order to reuse or recycle it after sorting [13, 20–21].

As one of the sources of solid waste, the cutting waste during denim jean production—which has relatively a homogeneous nature—is between 10% and 15%, and most of the waste is recycled by unraveling and reusing the fibers in the production of insert yarns (weft direction), which is a good example of "recycling in design (RiD)." Most jeans' producers recycle denim waste in their own manufacturing plants or have contracts with textile waste recyclers to reuse the waste material in the spinning of new yarns. There is also substantial trade of denim waste all over the world [13, 21–22].

Another source of denim waste is the post-consumer denim jeans. Color, quality of fabric, and garment accessories like rivets, buttons, zippers, and labels are the main components of the heterogeneous nature of this very denim waste. The main problem is the collection of post-consumer jeans. Although in many countries collection systems are in place, many consumers discard their jeans as solid municipal waste. Jeans that are collected are mostly sold to textile sorting companies which manually/automatically sort the rewearable jeans for sale to second-hand shops and in Third World countries. Nonwearable jeans are, however, shredded and used for the development of various types of products such as thermal and acoustic insulators and/or textile-based composites for certain structural and other specialized applications [13, 21, 23–28].

Finally, as a different approach to decreasing the denim waste regarding post-consumer denim jeans, the leasing of jeans was introduced several years ago by MUD Jeans. In this concept, the producer or distributor of the jeans stays the owner. The user of the jeans only "buys" the right to the use the jeans for a period of 1 year. If the jeans need to be repaired, it is at the cost of the lease company [22]. The same company, together with several of others, has also focused on "design for recycling (D4R)" for facilitating the recycling of denim jeans, such that they do not use leather labels but printed ones at the waistband, employ rivets and buttons that are made from 100% stainless steel and no finishing (electroplating), and have utilized buttons made out of recycled jeans on their denim shirts and sweaters, etc. [13, 29–30].

3. An overview of recycling in textiles and apparel

An average person buys 60% more clothing items every year and keeps them for half as long as they used to keep about 15 years ago. It is assumed that overall consumption of textiles will have reached to 102 million tons by 2030 and that the textile industry's waste will have increased by about 60% between 2015 and 2030. That means the total level of fashion waste will reach to 148 million tons in 2030. As is known, the majority of clothing waste ends up in landfills or is incinerated, and once in landfills, it takes hundreds of years for natural fibers to decompose and may release methane and CO_2 into the atmosphere. Synthetic materials, on the other hand, are not designed to decompose and may release toxic substances into groundwater and the surrounding soil. If the average life of clothing could be extended by only 3 months, it would reduce waste generation as well as their carbon and water footprints, by 5–10% [19, 31–34].

The textile industry's linear model of "make, use, and dispose of" represents an apparent pressure on scarce natural resources. Circular economy, on the other hand, aims to move away from the unsustainable linear model by decoupling economic activity from the consumption of finite resources and designing waste out of the system. When the recycling component is included, it helps to absorb the residuals of industrial and consumer use [35–36]. Accordingly, circular economy's principles may be given as follows:

- Put an end to waste generation and pollution during the design stage.
- Keep products and materials in use.
- Restore and regenerate natural systems.

Within that concept, a five-step waste management hierarchy was introduced in order to direct toward a more sustainable behavior (**Figure 1**) [37].

Waste generation prevention has the highest significance followed by reuse. Reusing is the concept of using undamaged parts of used products for manufacturing activities. When textiles turn into waste and are disposed by their consumers, recycling offers the opportunity to save raw materials and energy as well as to reduce pollution. Product/material recovery includes the activities like repairing, refurbishing, and disassembling, performed to regain the product value at the end of its life cycle. To dispose generated waste is the last step of the hierarchy [17, 38–41].

Textile recycling routes can be categorized in different ways as follows:

1. Mechanical, chemical, thermal, and biological based on the nature of the process.

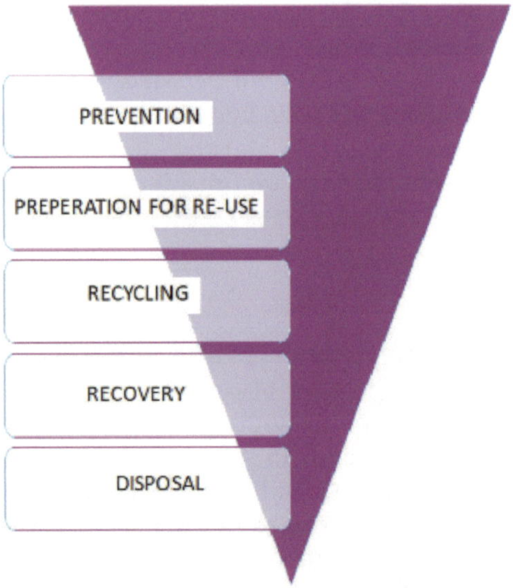

Figure 1.
Waste management hierarchy [38–39].

2. Downcycling where the product of the recycled material is of lower quality. As the length of fibers is shortened and the constituent of molecules is reduced by wear, laundry, and recycling process, the textile recycling is often in the way of downcycling.

3. Upcycling where the product of the recycled material is of higher quality.

4. Open-loop recycling covers a concept in which a product's raw material is separated to be utilized as a part of a randomly used item. Generally, the second item is not recycled and discarded toward the end of its life. Downcycling can be equalized with open-loop recycling system in which the reclaimed material is used to make a less valuable product than the disposed one. Therefore, it results in little economic value and low environmental benefits.

5. Closed-loop recycling refers to recycling techniques where the material recycled is a similar material which is being delivered. For example, the waste material reenters a piece of clothing production chain. Closed-loop recycling generates a greater impact on sustainability, and a product remains in a circular stream and retains its material quality. There are a number of ways to define closed-loop recycling approaches in the apparel industry. Three of them are [41]:

    - Recycling pre- or post-consumer textile waste.

    - Cradle-to-cradle (C2C) methodology in which waste is reclaimed and used again in the production of products of the same or higher value. Biological waste can be composted, while technical waste can be reused within industry to create the same products again.

    - Closed-loop reuse of existing garments: Although reuse of garments is not recycling in the sense of breaking down a product into its raw materials, the product may enter a new life cycle within the same production chain.

Textile recycling technologies are also categorized into four classes as primary, secondary, tertiary and, quaternary approaches [13, 40, 42–43]. Primary recycling involves recycling of material in its original form for recovery of equal value. Secondary recycling incorporates processing a post-consumer product into raw materials usually by mechanical means into a product with different physical and/or chemical properties (mechanical recycling). Tertiary includes processes like pyrolysis and hydrolysis, in which waste is converted to basic chemical constituents, monomers, or fuels (chemical recycling). Quaternary (recovery) covers waste-to-energy conversion processes such as incineration of solid waste or utilization of heat generated.

3.1 Textile recycling techniques

Recycling is the process of breaking down a product or material to make a material of a higher or equal value (upcycling) or of a lower value (downcycling), in which textiles are commonly mechanically or chemically broken down to their fiber constituents [41]. Biodegradation is another method used to recycle waste and to break down organic materials into compounds.

3.1.1 Mechanical recycling

The difference between mechanical and chemical recycling is that wet processing is eliminated or reduced in the mechanical recycling system [44]. Most of the current recycling systems for post-consumer waste textiles mainly include reuse and mechanical processes.

The method of mechanical recycling, which is categorized as a secondary recycling approach, is composed of two main processes: sorting of the waste material and the mechanical decomposition of the fabric. The material to be recycled is sorted according to fiber type, color, quality, etc. Sorting for post-industrial waste may be performed with a risk of uncertainty as the fiber content and properties of the fibers may not be always known. The in-house reprocessing of manufacturing-related waste represents recycling on the primary level [39].

The disintegration of textile material to a fibrous form through mechanical recycling is referred to as shredding or garneting [38, 40–41]. In mechanical recycling machines, the fabrics are cut into small pieces of 1 to 8 cm strips with a rotary blade and separated into single fibers through a process known as "picking," "pulling," or "tearing" by needle-equipped cylinders which have progressively smaller spiked surfaces. On such machines waste is fed through a conveyor belt of the front roller to be transferred to the spiked roller. Spiked roller rotates clockwise, and bottom roller, located under the spiked one, rotates anticlockwise. The distance between the rollers can be varied according to the type of input material, and the waste is opened while passing through rollers.

Mechanical recycling has some shortcomings since the process is too aggressive to retain fiber quality and can result in a 75% loss of value after the first cycle. The mechanical process breaks may cause a tremendous loss in fiber length and a significant decrease in the material quality. For the process, longer processing times are needed, and the production rate is lower. As a result, blending with virgin material (especially in the case of cotton and wool fibers) for spinning processes is inevitable [38, 42, 44]. Consistently, waste collected from the manufacturing supply chain produces higher-quality recycled fibers than those collected from post-consumer waste. The pre-consumer and post-industrial waste can be respun into yarns which are further woven or knitted into fabrics and then used in apparel, upholstery, etc. [13]. Heterogeneity of post-consumer waste worsens constant quality retention.

Nonetheless, despite the drawbacks of mechanical recycling, the technology has shown promising for the reprocessing of denim fabric and garments [45].

3.1.2 Chemical recycling

The method of chemical recycling, which is categorized as a tertiary recycling approach, involves chemical processing of the fiber polymers, e.g., depolymerizing or dissolving. Chemical recycling depends on the quality of the processed waste to a limited degree and decomposes fibers down to the polymeric level [39]. Various chemical recycling processes have been demonstrated and developed. Chemical recycling of synthetic polymers and feedstock recycling depolymerize waste plastics into base chemical molecule called monomers with high purity [38]. The presence of additives and chemicals used in the polymerization process affects the purity and quality of the monomers obtained after recycling. The thermochemical process used to decompose polymers is referred to as pyrolysis, sometimes thermolysis. Pyrolysis is conducted at various temperatures and pressure levels and with the presence of catalysts or reactive gases. Pyrolysis processes are only economically viable for certain manufactured fibers including polyesters, polyamides, and polyolefins [13, 40].

Mechanical recycling	Chemical recycling
Categorized as a secondary recycling approach	Categorized as a tertiary recycling approach
Wet processing is eliminated	Involves chemical processing
It is not as energy-intensive as chemical recycling	The biggest challenge is that chemical recycling is very energy-intensive
Mechanical recycling process is too aggressive to retain fiber quality	Chemical recycling allows a more valuable product in comparison to the products recycled by mechanical processes
Heterogeneity of post-consumer waste worsens constant quality retention	Chemical recycling depends on the quality of the processed waste to a limited degree
Mechanical recycling has been efficiently adopted by industry for recycling of single fiber materials	Chemical recycling is expected to be more suitable for large-scale recycling of blended materials

Table 1.
Comparison of mechanical and chemical textile recycling techniques.

Chemical recycling for polyester also includes glycolysis, hydrolysis, and metanolysis processes.

Chemical recycling is a promising process since it allows the recovery of a more valuable product in comparison to the products recycled by mechanical processes [34, 38, 42, 44]. As it uses a selective degradation method, chemical recycling is expected to be more suitable for large-scale recycling of blended materials, while mechanical recycling has been efficiently adopted by industry for recycling of single fiber materials. In products of cotton and polyester, the fibers can be chemically separated and then reformed into new fibers [13]. On the other hand, although chemical textile recycling has broader use than do the mechanical method, chemical and water consumption (70% lower in case of mechanical processing) for wet processing is high. Barriers to the widespread adoption of chemical recycling also include high costs, multiple processing steps requiring high operational knowledge, and energy requirement for heating and scouring processes [42, 44].

A comparison of mechanical and chemical textile recycling techniques are given in **Table 1** [38, 44].

3.1.3 Bioconversion processes

Biodegradation can be featured as a method used by nature to recycle waste and to break down organic materials into compounds by microorganisms such as bacteria, fungi, insects, worms, and others. Through biodegradation processes, it is possible for nature to clean up wastes, to provide nutrients for the growth of new lives, and to produce the energy necessary for various biological processes [46]. Biochemical transformation via fermentation is an attractive way for utilizing recycling textile waste. Cotton is typically composed of 88–96% cellulose, and it is possible to hydrolyze waste cotton by enzymatic or chemical methods to obtain glucose and then ferment it into value-added products. Biogas production from textile waste via anaerobic digestion is an alternative route to utilize solid waste from textile industry. Organic compounds in solid waste can be used as a raw material to produce desired products via bioconversion processes. On the contrary, thermal and chemical processes can convert both organic and inorganic compounds to value-added products [47].

3.1.4 Thermal and thermochemical conversion processes

Conventional thermal processing refers to the combustion of solid waste and its conversion into energy. Since solid waste from the textile industry contains a high energy content, it can be used as a raw material to generate heat energy [47]. Solid waste from the textile industry can be used as a raw material to produce briquette. The thermal processes that are performed at high temperature with inadequate oxygen could generate carbon monoxide, which is a greenhouse gas. Therefore, a thermochemical conversion process, such as pyrolysis, is applied. Pyrolysis is referred to the decompositional process with high temperature in the absence of oxygen condition [47]. Products from pyrolysis are various, such as activated carbon fiber, char, bio-oil, and syngas. The variation of product is related to pyrolysis condition.

3.2 Challenges in recycling denim

Despite its growing popularity, there are numerous obstacles to textile recycling [47–48]. The major ones to the optimization of textile recycling are:

1. Economic viability: Due to the widespread production of lower-grade products (downcycling) from textile recycling, many recycled textile wastes are unsuitable for multiple recirculation and use. Limited recirculation and reuse are not economically viable and discourages investment in textile recycling.

2. Composition of textile products: The base components of many textile products make them unsuitable for recycling. The presence of plastics and metals in textile products hinders their recyclability.

3. Nonavailability of recyclable textile materials: A limited quantity of used textiles and textile waste are collected and sorted for recycling, and the quantity that is suitable and accessible for recycling is insufficient.

4. Technological limitations: One of the main reasons for the limited quantity of recyclable materials is the lack of technologies for sorting textile waste in preparation for recycling. Dyes and other contaminants cannot be separated from the original fibers by most of the existing methods.

5. Lack of information and limited public participation: Limited public awareness on the merits of recycling contributes immensely to the low recycling rate, causing market inefficiency.

6. Poor coordination, weak policies, and standards: Uncoordinated collection of waste and absence of an integrated and well-coordinated framework and policies to enhance the overall efficiency of the textile recycling are identified as barriers to efficient recycling.

There are also some constraints and challenges faced specific to denim recycling processes [29, 49]. Collection and sorting of worn-out jeans is time-consuming and laborious. Labels; metal parts like rivets, zippers, and buttons; and leather patches have to be removed manually from the jean before shredding. Generally, the metal and leather parts are removed, but it is more difficult to remove the labels, and therefore jeans are sent along with them. The consequence is that the labels contaminate the recycled denim material as they are made of other materials. Any metal parts present on the jean to be recycled may cause problems to the machinery and process. It is easier to remove buttons and zippers by using gravitation but since rivets are too small and too light, special care is needed to remove them.

Denim jeans are characterized by thick lapped seams that create problems during shredding and carding processes. The presence of elastane is another problem. It is more convenient to separate it before shredding and cutting, but this can only be done by chemical recycling. Recycling different colored jeans together results in a multicolored yarn that can create problems in dyeing. Recycled fibers might not meet the quality of virgin ones and could not be spun or woven properly.

4. Life cycle assessment of a denim fabric

Life cycle assessment (LCA) is a methodology where the environmental performance of a product or service is assessed starting from the raw material extraction point to the end of life of that product/service, i.e., from "cradle to grave." The methodology of LCA is defined under the ISO 14040/44 Standard [50].

For a pair of denim jeans, the life cycle (**Figure 2**) starts with the production of raw materials such as fibers and chemicals. These materials are then transported to fabric manufacturer and processed to become a fabric. During fabric production, energy and water are consumed in addition to raw materials while emission to air and to water and production waste are generated. The following process for the fabric is garment manufacturing in which the fabric is cut, sewn, washed, and accessorized (rivets, buttons, etc.) according to the design. Finally, the finished garment is sent to a warehouse or directly stores to be sold. After it is bought, the garment is washed and dried (or dry-cleaned depending on its nature) many times throughout its use phase. When it completes its life span, the garment has various "end of life" scenarios such as recycling, reused, refurbished, and disposed in landfill or incinerated, etc., which were discussed in the previous sections.

Life cycle assessment (LCA) helps us analyze the environmental performance of denim production in a transparent and systematic way and identify the hot spots.

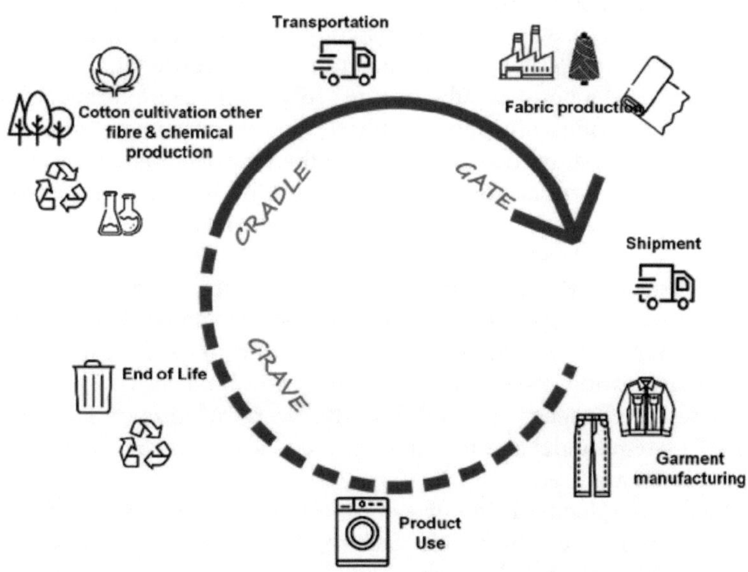

Figure 2.
Lifecycle of a jean.

4.1 Methodology used for assessing the lifecycle impacts

The methodology behind life cycle assessment (LCA) is to accumulate all the impacts originated from the inputs and outputs of a system starting from the "cradle to grave" and to give a total impact number for the system under discussion. Accordingly, in an attempt to exemplify the importance of recycling of denim from the perspective of its environmental impact, an experimental work regarding the life cycle assessment of a denim fabric with and without recycled fiber content was conducted. For the work, the inventory was based on the 2019 denim production figures of a Turkish denim manufacturer in their manufacturing plant in Turkey. As assessment tool, SimaPro software developed by the Pré Sustainability, was used. SimaPro is the leading life cycle assessment (LCA) software that has been used for more than 25 years by the industry and academics in more than 80 countries [51]. SimaPro uses two types of data: primary and secondary. The primary data involves the basics of a denim production such as the amount of cotton used to manufacture 1 m of denim, i.e., 0.5 kg cotton. This data is exclusive to the fabric production practices of the factory. The secondary data, however, comes from the database, and it includes the impacts originated from producing that much raw material (in our case cotton fiber) and all other inputs such as chemicals at every stage. For secondary data, Ecoinvent database that is embedded into the software and is the most common life cycle inventory (LCI) database worldwide is used [52].

To be able to perform life cycle assessment of a specific good or service, one needs to have inventory data for the complete supply chain. Due to the amount of data needed in order to be able to perform a life cycle assessment (LCA) study for a full supply chain, it is practically impossible to collect and organize the data of the complete background system. In that respect, the Ecoinvent database provides this very background system fulfilling the data required for the assessment. The Ecoinvent v3 database contains life cycle inventory (LCI) data taken from various sectors such as energy supply, agriculture, transport, building materials, production of chemicals, and metal production and consists of around 17,000 datasets, each of which describes a life cycle inventory (LCI) on a process level [52–53].

SimaPro software provides six libraries that each contain all the processes that are found in the Ecoinvent database but uses different system models and contains either unit or system processes. The three Ecoinvent system models are "allocation at point of substitution," "cutoff by classification," and "consequential." The system model "allocation at the point of substitution" follows the attributional approach in which burdens are attributed proportionally to specific processes. "Allocation, cut-off by classification" system model is based on the recycled content or cutoff approach in which the primary production of materials is always allocated to the primary user of a material. In this approach, if a material is recycled, the primary producer does not receive any credit for the provision of any recyclable materials available and burden-free to recycling processes; therefore recycled materials bear only the impacts of the recycling processes. The system model "substitution, consequential, long-term" uses different basic assumptions to assess the consequences of a change in an existing system and can be used for perspective studies and prediction of future changes [54]. In this study, for recycled materials "cutoff by classification" system model and for all other data "allocation at point of substitution" system model are used.

For life cycle assessment (LCA) of a product, the production of an item (e.g., denim fabric) is simulated, using both consumption and production data (primary) of a factory and the corresponding secondary data from the Ecoinvent database. The next step is to choose the environmental impacts to be calculated. The whole process is given in **Figure 3**.

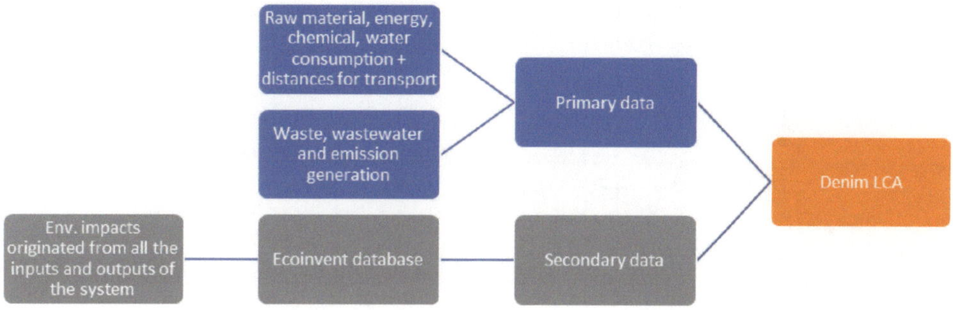

Figure 3.
LCA calculation process.

4.2 Selected impact categories

One of the most important parts of life cycle assessment (LCA) is the outputs, in other words the environmental impacts of the product. With SimaPro software, it is possible to calculate over 100 environmental impact categories. For that reason, the academic- and industry-specific life cycle assessment (LCA) studies' guidelines and standards were surveyed to determine the environmental impacts for our study (**Table 2**).

With this taxonomy (**Table 2**), the commonly used environmental impact categories are listed for textiles. Consequently, based on the taxonomy given in **Table 1** and the raw material need for a denim fabric, five impact categories are selected. These impact categories, their definitions, and calculation methodologies within the SimaPro software are presented in **Table 3**.

4.3 Environmental impact as a rationale for denim fabric recycling

4.3.1 System perspective

As was mentioned above, the life cycle assessment (LCA) methodology is selected to calculate the environmental impacts of denim fabrics having different recycled contents in the same article so that the whole system can be taken into consideration. This means that we have to calculate the effect of every step in the life cycle to see the whole impact of our choices, including cultivation/production of the fiber, fabric production, garment manufacturing, distribution, consumer laundering, reuse, and final disposal [76–77]. One may think that using recycled cotton reduces the impact drastically, but parameters such as increased waste during production and increased energy usage should also be taken into account in a system perspective.

If only one process or only fabric production as a system were calculated, this would have represented a single framed approach which is generally not preferable as calculations for production of single frames may lead to unwanted and unforeseen effects elsewhere in the whole system.

4.3.2 Five environmental impact categories

The specifications of the denim article selected for the work is given in **Table 4**. Life cycle assessment (LCA) was conducted for 1 m of the article in accordance with the process steps including fiber cultivation, transportation, and all the production steps covered in the Turkish denim manufacturing company. The five environmental impacts are presented in **Figure 4**. For the comparative study, the results are

References	Global warming potential (GWP) climate change	Acidification	Eutrophication	Ozone layer depletion	Abiotic depletion	Photochemical oxidant formation	Freshwater use	Human toxicity	Water consumption	Terrestrial ecotoxicity	Greenhouse gases (GHG)	Nonrenewable energy use	Carcinogens	Land occupation	Aquatic eutrophication	Mineral extraction	Ecotoxicity	Freshwater eutrophication	Freshwater aquatic ecotoxicity	Ionizing radiation	Water depletion
[55]	1	1	1	1			1		1												
[56]	1	1			1			1										1			1
[57]		1	1	1	1	1		1		1											
[58]											1	1									
[59]	1			1						1		1	1	1	1	1				1	
[60]		1	1			1					1		1				1				
[61]	1	1				1									1						
[62]	1	1	1						1			1									
[63]	1					1												1			
[64]	1	1	1	1									1	1		1	1			1	
[65]	1	1	1	1	1		1	1		1									1		
[66]											1										
[9]	1		1		1		1		1					1							
[67]	1		1		1																1
Total	10	8	8	5	5	4	3	3	3	3	3	3	3	3	2	2	2	2	2	2	2

Table 2.
A taxonomy of environmental impact categories for textiles.

Indicator	Unit	Description	Example impact	Methodology
Global warning potential	kg CO_2 eq (kilogram carbon dioxide equivalent)	Emission of greenhouse gases (GHGs)	Climate change	IPCC 2013 GWP 100a [68–69]
Freshwater use	lt (liters)	Excessive freshwater taken from the environment	Water scarcity	Life cycle inventory
Land use	m^2 a (meter square per annum)	The amount of agricultural area uccupied	Deforestation	ReCiPe 2016 Midpoint (H) [70–71]
Eutrophication potential (EP)	kg PO_4^{3-} eq (kilogram phosphate equivalent)	Emission of substances to water contributing to oxygen depletion	Nutrient loading to water stream-water pollution	CML-IA baseline [72–75]
Abiotic resource depletion	kg Sb eq (kilogram antimony equivalent)	Measure of mineral, metal, and fossil fuel resources used to produce a product	Mineral scarcity	CML-IA baseline [72]

Table 3.
Selected environmental impact categories.

Article Code	Width (cm)	Weight (oz/yd^2)	Composition
Article A	150	14.89	100% cotton

Table 4.
Article specifications.

given in terms of percentages (%) so that unit differences of the impact categories were eliminated.

In denim production, the hottest spot for the selected four categories is the fiber growth stage. In the fifth impact category, abiotic resource depletion, fiber stage has the second highest impact. This clearly shows the importance of raw material selection for denim fabric production.

4.3.3 Impact of recycled material content

This section aims to determine the impact of recycled cotton content in the denim fabric under discussion (**Table 4**). Accordingly, different recycled cotton contents are used in the life cycle assessment calculations of Article A. These are as follows:

- Article A-1: 100% cotton
- Article A-2: 80% cotton +20% recycled cotton
- Article A-3: 70% cotton +30% recycled cotton
- Article A-4: 60% cotton +40% recycled cotton
- Article A-5: 50% cotton +50% recycled cotton

Figure 4.
Five environmental impacts of selected fabric according to process steps.

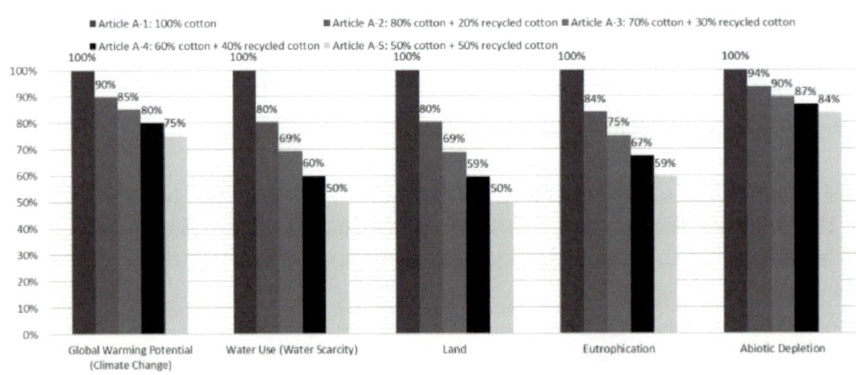

Figure 5.
Results of LCA calculations of virgin and post-consumer recycled cotton blends.

In each version, the recycled cotton content was increased by 10%. The recycled cotton used in the calculations is post-consumer recycled cotton, and its industrial data such as production and transportation data is obtained from a local supplier.

The results of the life cycle assessment (LCA) calculations are presented in **Figure 5**. As may be seen from these results, global warming potential decreases by

5%, eutrophication drops by 8%, and abiotic resource depletion drops by 3% with each addition of 10% recycled content in the blend.

Global warming potential, in other words climate change impact, is affected by two main stages: fiber cultivation and spinning. The energy usage in spinning increases when the recycled cotton content is increased, which implies a negative impact of the use of such fiber on global warming potential. However, since the percentage of the virgin cotton usage is decreased, this decline delivers a high positive impact for global warming potential, lessening the effect of energy usage.

In eutrophication calculation, the main effect derives from usage of fertilizers, pesticides, and insecticides at farm level. During irrigation of cotton, the probability of water pollution caused by these chemicals increases. Once the usage of virgin cotton decreases in the blend, the value of eutrophication decreases. Overall, the decrease results as 8% with a use of 10% recycled cotton.

Water use and land use impacts decrease by 10% with an addition of 10% recycled cotton. Since cotton uses land, and a high amount of water in the field during cultivation, avoiding the use of virgin cotton creates a high decrease in impact categories. If one can spin and weave a blend of 50% virgin and 50% recycled cotton, the overall impact on these two categories decreases 50%, which is a considerable figure when the amount of fabric produced reaches approximately 3 billion meters annually.

4.3.4 Impact of organic cotton and recycled cotton

This section aims to determine the impact of blending recycled cotton with organic cotton. Organic cotton data used in the calculations is generated from the literature [78–79]. The percentages used in the life cycle assessment (LCA) calculations are as follows:

- Article A-1: 100% cotton
- Article A-6: 100% organic cotton
- Article A-7: 80% organic cotton +20% recycled cotton

Once organic cotton is used, at least 25% decrease appears in three categories: namely, global warming potential, eutrophication, and abiotic resource depletion. The decrease in water use (11%) is comparably low. On the other hand, real decrease happens in the land use, nearly 40%. This is due to the data for Aegean Region organic cotton. The yield in Turkish organic cotton is comparably high. The land use for 1 kg of lint organic cotton is 4.65m^2 for Turkish organic cotton. The same figure appears to be 19.7 for global production and 20.9 for the US organic cotton (**Figure 6**).

The virgin cotton used in study for calculations is a blend of the US, Turkish, Greek, and Brazilian cotton.

The results of the life cycle assessment (LCA) calculations are presented in **Figure 7**. The data shows that diverting from 100% regular cotton to 100% organic cotton reduces global warming potential by 27%, eutrophication by 26%, abiotic depletion by 24%, and land use by 39%. In addition, as in Article A-7, blending organic cotton with 20% recycle cotton generates an additional 10% decrease in land use. With the aid of this blend, the comparably lowest impact in land use is achieved in this study. The same is true for global warming potential and abiotic depletion. Article A-7 has the lowest values compared to the rest of the articles in **Figures 5** and 7.

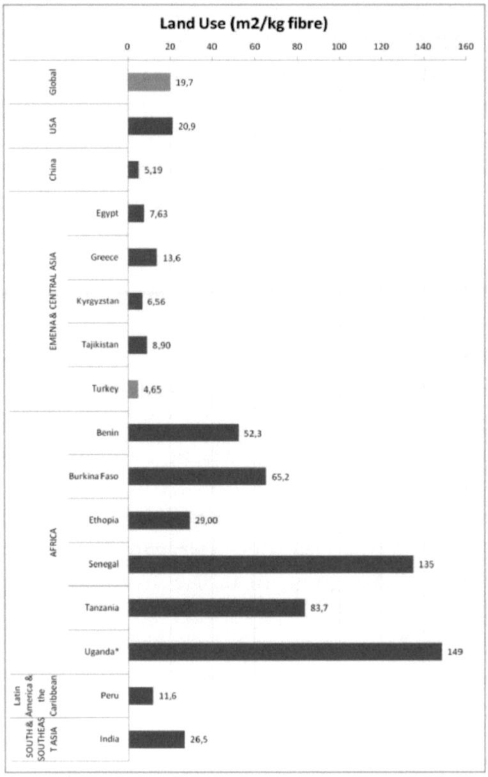

Figure 6.
Land needed for 1 kg lint organic cotton, m² [80].

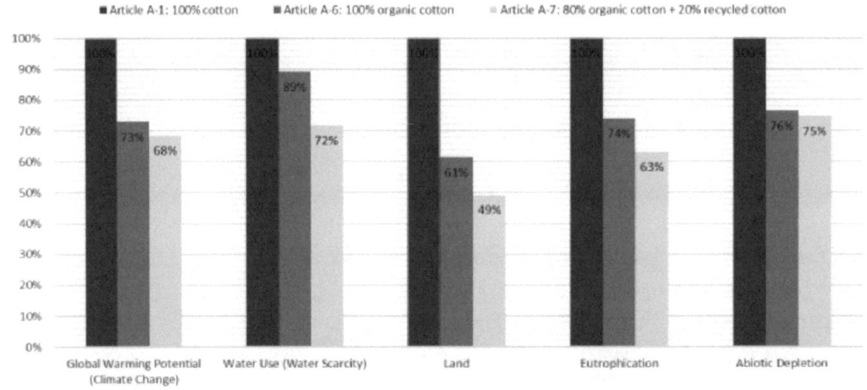

Figure 7.
Results of LCA calculations of organic and post-consumer recycled cotton blends.

On the other hand, eutrophication and water use remain higher than those of the articles in **Figure 5**. Organic cotton still uses a decent amount of water during cultivation. Therefore, Article A-7 could not result with the lowest impact on water use and eutrophication.

4.4 Discussion

Fiber selection stage—in this study fiber cultivation stage—has the main role in environmental impact of a denim fabric. Selected five impact categories are divided into the article's process steps starting from cradle, meaning cotton cultivation, to gate, the end of fabric production. When it is analyzed (Section 4.3.2), one can see that fiber has more than 50% of the total impact in four out of the five categories. Only in abiotic depletion, warp preparation and dyeing have a greater effect than that of the fiber itself. Since recycled cotton usage means eliminating the fiber growing stage or in other words not using virgin raw material that requires natural resource, the environmental impact decreases as the recycle content increases in the fabric. And this is mainly true for global warming potential, eutrophication, water use, and land use. Especially for water and land use, fiber growth stage has more than 90% impact on the overall score. Fifty percent recycled cotton use decreases both impacts by 50%. Therefore, it is better to use recycled content to decrease the environmental impact of water and land use mainly.

Organic cotton with recycled cotton combination has the lowest impact in land use. Using 100% organic cotton also helps reduce global warming potential and eutrophication more than using 50% recycled cotton. However, when it comes to water use and land use, recycled cotton always scores the best since it is not a grown raw material. Here, one can question the production of recycling. The data related to production and transportations are taken into account in the LCA calculations.

In conclusion, denim recycling is very crucial to reduce the water and land use impact of jeans. Combining recycled cotton with organic cotton also leads to reductions in other impact categories such as eutrophication, global warming potential, and abiotic depletion as well. As a future study, the impact of different fibers used in denim fabric production may also be analyzed with a combination of recycled cotton.

4.5 Challenges

Recycled cotton source appears to be the first challenge when whole textile system is considered. There are regulations in countries either limiting or declining the import of second-hand garments. This creates a limitation in source since the collection of second-hand garments is not organized in some countries.

The second challenge is the composition of jeans. Historically, the main composition was 100% cotton. Right now, more than half of the jeans include elastane fiber as well. Besides elastane, new compositions include man-made fibers and regenerated cellulosic fibers. The more complex the composition, the harder it gets to recycle jeans mechanically.

The most important challenge here is always the consumer mindset. Across the industry, only 13% of the total material input is in some way recycled after clothing use. Most of this recycling consists of cascading to other industries and is used in lower-value applications, for example, insulation materials, wiping cloths [81], Once discarded, over half the garments are not recycled but end up in mixed household waste and are subsequently sent to incinerators or landfill. According to a McKinsey analysis, as was mentioned before, an average consumer buys 60% more clothes per year than 15 years ago but keeps the clothes only half the time, and this really is shocking once the numbers become visible [82]. Consumer awareness should be increased via marketing channels and mainly in schools.

5. Future of denim recycling

Since mechanical recycling technique is a challenge in the process, recently, new techniques have emerged to use denim jeans and other cellulosic materials as a source/raw material. Companies like Re:newcell, Infinited Fibers, and Nanollose are taking second-hand garments, applying fiber separation and turning cellulosic part into liquid [83–85]. Some of them include fermentation, and as a last step, they turn the liquid into the material. The process resembles regenerated cellulosic fiber process. The use of fermentation appears to be a promising step into bio design for textiles, and this also eliminates all the negative sides of mechanical recycling.

Conflict of interest

The authors declare no conflict of interest.

Author details

Sedef Uncu Akı[1], Cevza Candan[2]*, Banu Nergis[2] and Neslihan Sebla Önder[1]

1 Orta Anadolu TAŞ, İstanbul, Turkey

2 Istanbul Technical University, İstanbul, Turkey

*Address all correspondence to: candance@itu.edu.tr

References

[1] Radhakrishnan S. The sustainable apparel coalition and the Higg index. In: Muthu SS, editor. Roadmap to Sustainable Textiles and Clothing: Regulatory Aspects and Sustainability Standards of Textiles and the Clothing Supply Chain. Singapore: Springer Science+Business Media; 2015. pp. 23-57. DOI: 10.1007/978-981-287-164-0

[2] Periyasamy AP, Militky J. Denim and consumers' phase of life cycle. In: Muthu SS, editor. Sustainability in Denim. 1st ed. United Kingdom: Woodhead Publishing, Elsevier Ltd; 2017. pp. 257-282. ISBN: 978-0-08-102044-9

[3] Sourcing Journal. Global Denim Industry to Grow by $14M by 2024, Thanks to One Innovation [Internet]. 2019. Available from: https://sourcingjournal.com/denim/denim-business/technavio-report-denim-organic-cotton-market-growth-182319/ [Accessed: 23 March 2020]

[4] Mordor Intelligence. Jeans Market - Growth, Trends, and Forecast (2020–2025) [Internet]. 2019. Available from: https://www.mordorintelligence.com/industry-reports/jeans-market [Accessed: 23 March 2020]

[5] Statista. Market size of the global denim market 2019–2023 [Internet]. 2019. Available from: https://www.statista.com/statistics/877056/denim-market-value-worldwide/ [Accessed: 23 March 2020]

[6] Grand View Research. Denim Jeans Market Size, Share & Trends Analysis Report by End User (Children, Men, Women), by Sales Channel (Offline, Online), by Region (North America, APAC, Europe, MEA), and Segment Forecasts, 2019–2025 [Internet]. 2019. Available from: https://www.grandviewresearch.com/industry-analysis/denim-jeans-market [Accessed: 23 March 2020]

[7] Textilengence. Global Denim Fabric Trade Continues Growth Trend. 2019. Available from: https://www.textilegence.com/en/global-denim-fabric-trade-continues-growth [Accessed: 23 March 2020]

[8] Fibre2Fashion. Denim companies answer to the call for sustainable jeans [Internet]. 2014. Available from: https://www.fibre2fashion.com/industry-article/7245/denim-companies-answer-to-the-call-for-sustainable-jeans [Accessed: 23 March 2020]

[9] Levi Strauss & Co [report]. The Life Cycle of a Jean - Understanding the environmental impact of a pair of Levi's 501 ® jeans. San Francisco. 2015. Available from: https://docplayer.net/411176-The-life-cycle-of-a-jean-understanding-the-environmental-impact-of-a-pair-of-levi-s-501-jeans-levi-strauss-co-2015.html [Accessed: 14 April 2020]

[10] Mahmud A. Environmental issues of denim processing [Internet]. 2019. Available from: https://textilelearner.blogspot.com/2019/02/environmental-issues-denim-processing.html [Accessed: 30 April 2020]

[11] Khan MR, Mondal IH, Uddin Z. Sustainable washing for denim garments by enzymatic treatment. Journal of Chemical Engineering. 2012; **27**:27-31. DOI: 10.3329/jce.v27i1.15854

[12] Periyasamy AP, Wiener J, Militky J. Life-cycle assessment of denim. In: Muthu SS, editor. Sustainability in Denim. 1st ed. United Kingdom: Woodhead Publishing: Elsevier Ltd; 2017. pp. 83-110. ISBN: 978-0-08-102044-9

[13] Leonas KK. The use of recycled fibers in fashion and home products. In:

Muthu SS, editor. Textiles and Clothing Sustainability: Recycled and Upcycled Textiles and Fashion. 1st ed. Singapore: Springer; 2017. pp. 55-75. DOI: 10.1007/978-981-10-2146-6

[14] Uniqlo. Feature: Making more sustainable jeans – Jeans Innovation Center [Internet]. 2020. Available from: http://www.uniqlo.com/en/sustainability/planet/special/jeans/ [Accessed: 23 March 2020]

[15] MLF G, JCC A, Bressers H, editors. Towards Zero Waste-Circular Economy Boost, Waste to Resources. 1st ed. Switzerland AG: Springer Nature; 2019. p. 273. DOI: 10.1007/978-3-319-92931-6

[16] Muthu SS, editor. Circular Economy in Textiles and Apparel. 1st ed. United Kingdom: Woodhead Publishing, Elsevier Ltd; 2019. p. 246. DOI: 10.1016/C2017-0-03221-4

[17] Shirvanimoghaddam K, Motamed B, Ramakrishna S, Naebe M. Death by waste: Fashion and textile circular economy case. The Science of the Total Environment. 2020;**718**: 137317. DOI: 10.1016/j.scitotenv.2020.137317

[18] Nellström M, Sariclca M. Comparative life cycle assessment of nudie jeans' repair and reuse concept [thesis]. Gothenburg: Chalmers University of Technology; 2019

[19] Yousef S, Tatariants M, Tichonovas M, Sarwar Z, Jonuškienė I, Kliucininkas L. A new strategy for using textile waste as a sustainable source of recovered cotton. Resources, Conservation and Recycling. 2019;**145**:359-369. DOI: 10.1016/j.resconrec.2019.02.031

[20] Yousef S, Tatariants M, Tichonovas M, Kliucininkas L, Lukosiute SI, Yan L. Sustainable green technology for recovery of cotton fibers and polyester from textile waste. Journal of Cleaner Production. 2020;**254**:120078. DOI: 10.1016/j.jclepro.2020.120078

[21] Manish R, Shahab F, Naresh T. Recycled materials as a potential replacement to synthetic sound absorbers: A study on denim shoddy and waste jute fibers. Applied Acoustics. 2020;**159**:107070. DOI: 10.1016/j.apacoust.2019.107070

[22] Luiken A, Bouwhuis G. Recovery and recycling of denim waste. In: Paul R, editor. Denim-Manufacture, Finishing and Applications. 1st ed. United Kingdom: Woodhead Publishing Series in Textiles, Elsevier Ltd; 2015. pp. 527-540. ISBN 978-0-85709-849-8

[23] Al-Sabaeei A, Napiah M, Sutanto M, Alaloul W. Effects of waste denim fiber (WDF) on the physical and rheological properties of bitumen. IOP Conference Series: Materials Science and Engineering. 2019;**527**:012047527. DOI: 10.1088/1757-899X/527/1/012047

[24] Baghaei B, Temmink R, Skrifvars M. Recycling of end-of-life textile materials by fabrication of green composites. 21st International Conference on Composite Materials (ICCM-21): 20-25th August 2017, Xi'an: China. Available from: http://www.iccm-central.org/Proceedings/ICCM21proceedings/papers/3451.pdf [Accessed: 14 April 2020]

[25] Ozkur S, Sezgin H, Akay E, Enis IY. Hybrid bio-based composites from blends of epoxy and soybean oil resins reinforced with jute woven fabrics. Materials Research Express. 2020;**7**: 015335. DOI: 10.1088/2053-1591/ab6892

[26] Temmink R, Baghaei B, Skrifvars M. Development of bio-composites from denim waste and thermoset bio-resins for structural applications. Composites: Part A. 2018;**106**:59-69. DOI: 10.1016/j.compositesa.2017.12.011

[27] Zeng B, Wang X, Byrne N. Development of cellulose based aerogel utilizing waste denim-a morphology study. Carbohydrate Polymers. 2019;**205**: 1-7. DOI: 10.1016/j.carbpol.2018.09.070

[28] Silva TL, Cazetta AL, Souza PSC, Zhang T, Asefa T, Almeida VC. Mesoporous activated carbon fibers synthesized from denim fabric waste: Efficient adsorbents for removal of textile dye from aqueous solutions. Journal of Cleaner Production. 2018;**171**: 482-490

[29] Radhakrishnan S. Denim recycling. In: Muthu SS, editor. Textiles and Clothing Sustainability - Recycled and Upcycled Textiles and Fashion. Textile Science and Clothing Technology. Singapore: Springer Science+Business Media; 2017. pp. 79-85. DOI: 10.1007/978-981-10-2146-6

[30] MUD jeans. 2020. Available from: https://mudjeans.eu/blog-denim-school [Accessed: 23 March 2020]

[31] Global Fashion Agenda & The Boston Consulting Group. The Pulse of the Fashion Industry [Internet]. 2017. Available from: http://globalfashionagenda.com/wp-content/uploads/2017/05/Pulse-of-the-Fashion-Industry_2017.pdf [Accessed: 11 March 2020]

[32] The European market potential for recycled fashion, CBI Ministry of Foreign Affairs [Internet]. Available from: https://www.cbi.eu/node/3065/pdf/ [Accessed: 17 March 2020]

[33] Textile and Garment Recycling Facts and Figures by Rick Reblanc [Internet]. 2019. Available from: https://www.thebalancesmb.com/textile-recycling-facts-and-figures-2878122 [Accessed: 13 March 2020]

[34] Rengel A. Be Sustainable. Recycled Textile Fibers and Textile Recycling-Report. [Internet]. 2017. Available from: study-on-recycled-textiles-and-textile-recyclability-ch.pdf [Accessed: 13 March 2020]

[35] What is a circular economy? A framework for an economy that is restorative and regenerative by design [Internet]. Available from: https://www.ellenmacarthurfoundation.org/circular-economy/concept [Accessed: 13 March 2020]

[36] Textile Recycling Toolbox [Internet]. 2018. Available from: https://globalfashionagenda.com/publications/#2020commitmenttoolboxes [Accessed: 13 March 2020]

[37] Being wise with waste: The EU's approach to waste management [Internet]. 2010. Available from: https://ec.europa.eu/environment/waste/pdf/WASTE%20BROCHURE.pdf [Accessed: 15 March 2020]

[38] Björquist S. Chemical recycling of cotton and polyester textiles [thesis]. The Swedish School of Textiles: University of Boras; 2017

[39] Neuß J, Schlich M. Best Before: A selective service life analysis of denim fabrics with a focus on washing and drying degradation to optimize their recycling efficiency [thesis]. The Swedish School of Textiles: University of Boras; 2019

[40] Muthu SS, editor. Sustainable Innovations in Recycled Textiles. Singapore: Springer; 2018. p. 82. DOI: 10.1007/978-981-10-8515-4

[41] Payne A. Open and closed loop recycling of textile and apparel products. In: Muthu SS, editor. Handbook of Life Cycle Assessment (LCA) of Textiles and Clothing. United Kingdom: Woodhead Publishing Series in Textiles, Elsevier Ltd; 2015. pp. 103-121. ISBN: 9780081001875

[42] Le K. Textile Recycling Technologies, Coloring and Finishing Methods [report]. UBC Sustainability Scholar. 2018. Available from: https://sustain.ubc.ca/sites/sustain.ubc.ca/files/Sustainability%20Scholars/2018_Sustainability_Scholars/Reports/2018-25%20Textile%20Recycling%20Technologie

s%2C%20Colouring%20and%20Finishing%20Methods_Le.pdf [Accessed: 16 March 2020]

[43] Wang Y. Fiber and textile waste utilization. Waste and Biomass Valorization. 2020;**1**:135-143. DOI: 10.1007/s12649-009-9005-y 1(1)

[44] Yavari R. Analysis of a garment-oriented textile recycling system via simulation approach [thesis]. Industrial Engineering Graduate Program-University of Windsor; 2019

[45] Gordon S, Hsieh Y. Cotton Science and Technology. England: Woodhead Publishing in Textiles; 2007. p. 490. ISBN-10: 1-84569-026-5

[46] Muthu SS, editor. Roadmap to Sustainable Textiles and Clothing-Environmental and Social Aspects of Textiles and Clothing Supply Chain. 1st ed. Singapore: Springer; 2014. p. 287. DOI: 10.1007/978-981-287-110-7

[47] Pensupa N, Leu SY, Hu Y, Du C, Liu H, Jing H, et al. Recent trends in sustainable textile waste recycling methods: Current situation and future prospects. Topics in Current Chemistry. 2017;**375**:76. DOI: 10.1007/s41061-017-0165-0-Bio conversion

[48] Filho WL, Ellams D, Han S, TylerD BVJ, Paço A, Moora H, et al. Review of the socio-economic advantages of textile recycling. Journal of Cleaner Production. 2019;**218**:10-20. DOI: 10.1016/j.jclepro.2019.01.210

[49] Heerink M, Koenen A. Experiences and Bottlenecks with Recycling of Jeans [Student Research Project]. Netherlands: Sustainable International Business: Saxion University of Applied Sciences; 2013

[50] DIN EN ISO 14040:2006–10, Environmental management - Life cycle assessment - Principles and framework (ISO 14040:2006) and Requirements and guidelines

[51] SimaPro LCA Package. Pré Sustainability [licensed software]. 2020. Available from: https://www.pre-sustainability.com [Accessed: 13 April 2020]

[52] Ecoinvent Centre. 2020. Available from: https://www.Eco-invent.org [Accessed: 16 March 2020]

[53] Lia H, Denga Q, Zhang J, Xi B, Skitmore M. Assessing the life cycle CO_2 emissions of reinforced concrete structures: Four cases from China. Journal of Cleaner Production. 2019; **201**:1496-1506. DOI: 10.1016/j.jclepro.2018.11.102

[54] Wernet G, Bauer C, Steubing B, Reinhard J, Moreno-Ruiz E, Weidema B. The ecoinvent database version 3 (part I): Overview and methodology. International Journal of Life Cycle Assessment. 2016;**21**:1218-1230. DOI: 10.1007/s11367-016-1087-8

[55] Cotton Incorporated [report]. Life Cycle Assessment of Cotton Fiber & Fabric Full Report. USA. 2012. Available from: http://resource.cottoninc.com/LCA/LCA-Full-Report.pdf [Accessed: 13 April 2020]

[56] Cardos AAM. Life cycle assessment of two textile products wool and cotton [thesis]. Porto: Universidade de Porto; 2013

[57] Shen L, Patel M. Life cycle assessment of man-made cellulose fibers. Lenzinger Berichte. 2010;**88**:1-59

[58] Shen L, Patel M. Life cycle assessment of polysaccharide materials: A review. Journal of Polymers and the Environment. 2008;**16**:154-167. DOI: 10.1007/s10924-008-0092-9

[59] Murugesh KB, Selvadass M. Life cycle assessment for the dyeing and finishing process of organic cotton knitted fabrics. Journal of Textile and Apparel, Technology and Management. 2013;**8**:1-16

[60] Vogtlander J. LCA benchmarking study on textiles made of cotton, polyester, nylon, acryl, or elastane. International Journal of Life Cycle Assessment. 2014;**19**:331-356. DOI: 10.1007/s11367-013-0626-9

[61] Baydar G, Filiz N, Mammadov A. Life cycle assessment of cotton textile products in Turkey. Resources, Conservation and Recycling. 2015;**104** (Part A):213-223. DOI: 10.1016/j.resconrec.2015.08.007

[62] Kalliala E, Nousianien P. Life cycle assessment environmental profile of cotton and polyester-cotton fabrics. Autex Research Journal. 1999;**1**:8-20. Available from: http://www.autex.org/v1n1/2264_99.pdf [Accessed: 13 April 2020]

[63] Manda BMK, Worrell E, Patel M. Prospective life cycle assessment of an antibacterial T-shirt and supporting business decisions to create value. Resources, Conservation and Recycling. 2015;**103**:47-57. DOI: 10.1016/j.resconrec.2015.07.010

[64] Bevilacqua M, Ciarapica F, Mazutto G, Paciarotti C. Environmental analysis of a cotton yarn supply chain. Journal of Cleaner Production. 2014;**82**: 154-165. DOI: 10.1016/j.jclepro.2014.06.082

[65] Ullah A, Perret S, Gheewala S, Soni P. Eco-efficiency of cotton-cropping systems in Pakistan: An integrated approach of life cycle assessment and data envelopment analysis. Journal of Cleaner Production. 2015;**134**:623-632. DOI: 10.1016/j.jclepro.2015.10.112

[66] Wang C, Lihua W, Xinlei L, Chong D, Ding D, Jia J, et al. Carbon footprint of textile throughout its life cycle: A case study of Chinese cotton shirts. Journal of Cleaner Production. 2015;**108**:464-475. DOI: 10.1016/j.jclepro.2015.05.127

[67] Sustainable Apparel Coalition [report]. Higg Materials Sustainability Index (MSI) Methodology. 2019. Available from: https://msi.higg.org/uploads/msi.higg.org/sac_textpage_section_files/27/file/MSI_Methodology_8-6-19.pdf [Accessed: 15 April 2020]

[68] IPCC. GWP 100a: methodology developed by the Intergovernmental Panel on Climate Change. 2013. Available from: http://www.ipcc.ch/ [Accessed: 13 April 2020]

[69] Climate Change 2013: The Physical Science Basis [internet]. Working Group I contribution to the Fifth Assessment Report of the IPCC. 2016. Available from: http://www.climatechange2013.org [Accessed: April 13, 2020]

[70] ReCiPe 2016 v1.1. A harmonized life cycle impact assessment method at midpoint and endpoint level Report I: Characterization. RIVM Report 2016-0104a [Internet]. 2017. Available from: http://www.rivm.nl/en/Topics/L/Life_Cycle_Assessment_LCA/Downloads/Documents_ReCiPe2017/Report_ReCiPe_Update_2017 [Accessed: 13 April 2020]

[71] Huijbregts MAJ, Steinmann ZJN, Elshout PMF, Stam G, Verones F, Vieira M, et al. ReCiPe2016: A harmonised life cycle impact assessment method at midpoint and endpoint level. International Journal of Life Cycle Assessment. 2017;**22**:138-147. DOI: 10.1007/s11367-016-1246-y

[72] CML-IA baseline methodology (version 3.05). The Netherlands; Leiden University. 2017. Available from: http://cml.leiden.edu/software/data-cmlia.html [Accessed: 13 April 2020]

[73] Guinee JB, Gorree M, Heijungs R, Huppes G, Kleijn R, van Oers L, et al. Handbook on Life Cycle Assessment. Operational Guide to the ISO Standards; Netherlands, 2001. p. 1, 2a, 2b, 3

[74] van Oers L, Huppes G. LCA normalisation factors for the Netherlands, Western Europe and the World. International Journal of Life Cycle Assessment. 2001;**6**:256. DOI: 10.1007/BF02978785

[75] Sleeswijk AW, van Oers L, Guinee J, Struijs J, Huijbregts M. Normalisation in product life cycle assessment: An LCA of the global and European economic systems in the year 2000. Science of the Total Environment. 2008;**390**(1): 227-240. DOI: 10.1016/j.scitotenv.2007.09.040

[76] The International EPD System [report]. Product Category Rules for creating Environmental Product Declaration (EPD) for Woven, Knitted and Crocheted Fabrics of Natural Fibers (Except Silk) for Apparel Industry. 2018. Available from: https://www.environdec.com/PCR/Detail/?Pcr=13246 [Accessed: 14 April 2020]

[77] Fletcher K. Sustainable Fashion and Textiles: Design Journey. 1st ed. Earthscan: UK; 2008. p. 254

[78] LCA of organic cotton Fiber - A Global Average. 2020. Available from: https://textileexchange.org/downloads/life-cycle-assessment-of-organic-cotton/ [Accessed: 13 April 2020]

[79] T.C. Tarım ve Orman Bakanlığı, Pamuk Araştırma Enstitüsü Müdürlüğü. 2020. Available from: https://arastirma.tarim.gov.tr/pamuk/Sayfalar/AnaSayfa.aspx [Accessed: 13 April 2020]

[80] Organic Cotton Market Report [Internet]. 2019. Available from: https://store.textileexchange.org/product/2019-organic-cotton-market-report/ [Accessed: 13 April 2020]

[81] A new textiles economy: Redesigning fashion's future. Ellen MacArthur Foundation [Internet]. 2017. Available from: http://www.ellenmacarthurfoundation.org/publications [Accessed: 13 April 2020]

[82] Pulse of the fashion industry [Internet]. 2017. Available from: https://www.commonobjective.co/article/pulse-of-the-fashion-industry-2017 [Accessed: 13 April 2020]

[83] Re:newcell. 2020. Available from: https://renewcell.com/ [Accessed: 15 April 2020]

[84] Infinite fiber. 2020. Available from: https://infinitedfiber.com/ [Accessed: 15 April 2020]

[85] Nanolose. 2020. Available from: https://nanollose.com/ [Accessed: 15 April 2020]

Chapter 5

Investigation of Sound Absorption Characteristics of Textile Materials Produced from Recycled Fibers

Nilgün Özdil, Gonca Özçelik Kayseri and Gamze Süpüren Mengüç

Abstract

Excessive sound causes growing public well-being problems and significant environmental contamination in our daily life. Generally, most of the noise problems are difficult to be treated at source, and the reduction of noise emission is usually achieved through the use of noise isolation processes. In recent times, nonwovens as one of the most common textile products have become valuable sound absorption materials. These materials are used as sound absorbers, sound diffusers, noise barriers, and sound reflectors. For sustainable development of the textile industry, solutions for both decreasing waste and reducing noise have been searched for years. Due to the good sound absorbing properties, recycled materials are becoming an attractive option to traditional materials for practical purposes. In this study, the measurement methods of acoustic characteristics of textile materials are explained, and the sound absorption features of nonwoven fabrics made from both pure and recycled polyester and polypropylene fibers are compared.

Keywords: sound absorption, acoustic textiles, textile material, sustainability, recycled fibers

1. Introduction

With the rising number of new technologies, daily noise pollution is becoming a growing public health problem along with the severe environment pollution affecting the life comfort, especially in big cities. The noise being exposed for a long time makes communication difficult, distorts concentration, and lowers quality of life, that is, health hazard. The World Health Organization (WHO) defines noise pollution as a worldwide problem influencing the health of numerous individuals. The most common reactions are headaches, fatigue, tension, body aches, sleep disturbances, and other problems related to stress. There are also studies showing the relationship between high blood pressure, vascular spasms, and heart attack associated with prolonged exposure to noise [1–3].

It is crucial to have appropriate living conditions for the people staying at homes and the workers to work well in the industries. Hence, in order to eliminate the problems faced by the people, the noise pollution needs to be decreased or removed. Transportation systems, machines, and constructions are the causes for the outdoor noise pollution, while interior noise can be produced by means of inside actions, performances, and operating electrical-powered equipment. Both outdoor and indoor noise pollutions are damaging for the people and animals. According to the ISO 25267 standard, sound pressure levels in living room, kitchen, and bedroom are recommended to be lower than 30, 40, and 20 dB, respectively. Subjecting to a noise level of 85 dB and higher for a long time can lead to increased blood pressure, raised stress level, and long-lasting hearing injury [4, 5].

Materials used for controlling the noise can be generally categorized as those appropriate for treatment at source, those for declining along the noise path, and those for receiver treatment. The classification can be done by absorbing materials, scattering or altering the sound energy on impact; barrier materials, blocking the sound waves; vibration isolation, damping the materials for reducing radiated sound; and silencers and passive and active materials, quelling the sounds [6].

Generally, it is not possible to handle most of the noise problems at source, and the decrease in noise emission is mostly undertaken by noise isolation systems. Active and passive noise control methods are available for noise control. Active noise control is the total noise reduction process using a secondary audio resource, while passive noise control is supplied by sound absorbers and sound insulation materials. Active sound control is managed by using complex digital signal processing techniques to delay the harmful sound waves [7]. Sound absorption is made in passive control method, and knowing acoustic properties of these absorbent materials is helpful for determining the suitable usage fields of the products such as sound barriers, walls, and road surfaces [8].

The performance of insulation materials should be evaluated in terms of multiscale principles, respecting physical features, protection level of health and environment, and applicability as building elements and their cost. The basic requirements on insulating products are good thermal and sound insulation characteristics. Noise reduction coefficient of 0.5–0.9 and thermal conductivity coefficient of 0.030–0.045 W/mK are the main characteristics of the most conventional insulating materials [9].

Nonwoven textile materials are widely used for sound insulation and absorption applications due to their porous structure in recent times. These materials are used as sound absorbers, sound diffusers, noise barriers, and sound reflectors. For the sound transmission through friction, the sound wave should penetrate into the absorbent material. The loss of sound energy in textile materials is influenced by various physical issues such as fiber type, fiber diameter, material thickness, density, bonding method, air resistance, and porosity. The nonwoven surface forms a large inner surface in the form of a flexible fiber skeleton of a fabric composition, providing these required physical conditions [3, 10, 11].

For sustainable development of the textile industry, solutions for both decreasing waste and reducing noise have been searched for years. Caused by the reduction of ecological supplies and the rising costs of waste removal, the recycling or reprocessing of textile waste is becoming progressively significant all over the world in terms of economy, environment, and sustainability [12, 13]. Both natural and synthetic fibers together with recycled materials are used to construct textile-based acoustic materials. The use of textile surfaces produced from recycled materials has spread to a wide range of areas, and its use as sound insulation material has also become widespread.

In this study, recycling nonwoven materials are considered in terms of sound insulation characteristics; the measurement methods of acoustic features of textile materials are explained; and the sound absorption properties of nonwoven fabrics produced from both pure and recycled polyester and polypropylene fibers are compared.

2. Recycling of textile materials

The growth of worldwide population, the rapid development of technology, the change of living standards and fashion trend, the improvement in marketing activities, and reduced time that the clothing is used before disposal result in serious increment in textile production and consumption. Along with these key factors for the rapid development in the textile industry, an overconsumption of clothing and an extreme use of resources and energy have started to become crucial problems and significantly responsible for the generation of textile waste as well.

In order to reuse the textile articles, the end-of-life textiles should be collected separately from the remaining waste, and reusable and nonreusable clothing have to be separated from each other. Textiles intended to be used in the second-hand market should be cleaned, dried, and not worn. According to Directive 2008/98/EC (European Parliament and Council of the European Union, 2008), recycling means any recovery action in which waste materials are reprocessed into products, materials, or substances for the original or other purposes. This description separates recycling from reuse required preparation by the word reprocess. Recovery, on the other hand, denotes any operation, either processing waste by replacing materials used to fulfill a function, or replacing waste being prepared to fulfill that function, in the plant or in the wider economy [14, 15].

Textile wastes released in and after the production of textile products have a large volume, and their disposals have become one of the most important threats to the ecological system. Apart from this, the raw materials used in this process are natural and limited resources; high water consumption and water pollution from this process as well as high energy consumption and the resulting carbon emissions have exceeded 1.2 billion tons per year [16]. Shortage of natural resources and increase of raw material cost have increased the importance of recycling. Due to the environmental troubles, the uses of biodegradable and recyclable material in many products become significant [5]. Theoretically, 97% of textile waste can be recycled [9]. In addition, waste storage areas are decreasing, and all these factors have created a serious threat to the environment and increased awareness in global world. As a result, the concepts of "recycling" and "sustainability" have gained great importance in the textile industry and have been among the priority state policies in the world.

Recycling in textiles can be defined as using the waste materials as raw resources to produce new products. According to the production flow of the new product produced by recycling textile wastes, recycling is evaluated in two groups such as open loop recycling and closed loop recycling. Open loop recycling characterizes the usage of recycled material in different purposes from the used material, whereas the closed loop recycling is the production of the used materials in the same product [17].

Textile recycling material can be also categorized into two groups such as pre-consumer and post-consumer wastes. Pre-consumer waste contains materials from the various production stages of textile industries, and these wastes are remanufactured for the automotive, aeronautic, home building, furniture, mattress, coarse yarn, home furnishings, paper, apparel, and other industries. Post-consumer waste is described as any type of discarded garment or household articles, which are worn out, damaged, outgrown, or have gone out of fashion. The post-consumer textiles are "opened" mechanically or chemically so as to return it to a fibrous form, and they can be able to process new products for renewed consumption [18].

Textile recycling technologies are realized in four different processes. Primary recycling is the recycling of a product to its initial state. In secondary recycling, waste is included in a different application area than its original form and transformed into a new product having lower levels of physical, mechanical, and chemical properties. In tertiary recycling, pyrolysis occurs by turning gas into simple chemicals or fuels by gasification and hydrolysis. The fourth recycling process is realized by utilizing the heat released by incineration of solid wastes. The recycling method of the fibers differs on the type of fibers, for instance, synthetic fibers are chemically recycled, whereas all others are mechanically recycled [19]. Recycling of synthetic fibers such as polyester and polyamide is included in primary recycling technologies and is evaluated within the scope of the "closed loop recycling" class. Today, the most applied recycling is "open loop recycling," which belongs to the secondary recycling technology class [13, 20].

Since recycled materials demonstrate good sound absorbing characteristics, these materials are becoming an interesting alternative to typical materials for functional applications. Acoustic barriers and acoustic ceilings, passenger vehicle noise absorbers, and wall claddings are some of the implementation of noise control functions of the nonwoven fabrics. Nonwoven fabrics made from recycled fibers are having more advantageous in terms of environmental pleasantness compared to traditionally utilized polyurethane foams produced by environmental harmful manufacturing methods and cannot be recycled [21]. Lower product cost, good processing, and environmental protection are some of the benefits of using recycled polyester nonwovens compared to conventional sound absorbers. Former studies about the noise absorption of nonwovens have indicated that the noise absorption coefficients of these materials in the high frequency range (f > 2000 Hz) are equivalent to that of the traditional sound absorbers such as rockwool and glass fiber [22].

3. Sound absorption characteristics of textile materials

3.1 Acoustic textiles

The definition of sound is a pressure change in air, water, or similar elastic medium, which can be perceived by the ear as a stimulus of hearing. Loudness and tone are two aspects of sound. Loudness is defined as the sound pressure expressed in decibel (dB), and tone is defined as the sound frequency expressed in Hertz (Hz). Human ear is not sensitive to all the sound frequencies, and depending on its frequency and intensity, the sound may or may not be audible to the human ear. The range of 20–20,000 Hz is the audible sound frequency for humans [4].

As the material or item is vibrated, the sound is created. These vibrations spread in solid, liquid, or gas medium in a waveform from the emitter to the receiver. Therefore, a sound wave is the transfer of energy radiated by a source material or an object into the medium as it travels. A sound wave is considered by its frequency, wavelength, and amplitude. The wave properties can change according to the interaction of sound waves with the receiver's surface and the properties of the receiver's object. The sound wave can be absorbed, transmitted, reflected, refracted, and diffracted from the surface [5].

The classic solution to the problem of noise pollution is to remove the noise from the source; however, it is not always possible to remove it in this way. Thus, noise isolation methods are usually used for the reduction of noise emission. For this purpose, two types of methods are used, defined as sound absorption and sound insulation, as shown in **Figure 1**. In sound absorption, air particles rub inside the insulating material, and with the conversion of the sound kinetic energy into heat energy, the sound power is reduced. Therefore, sound absorption refers to the absorption level of the sound from the sound source in the environment of the

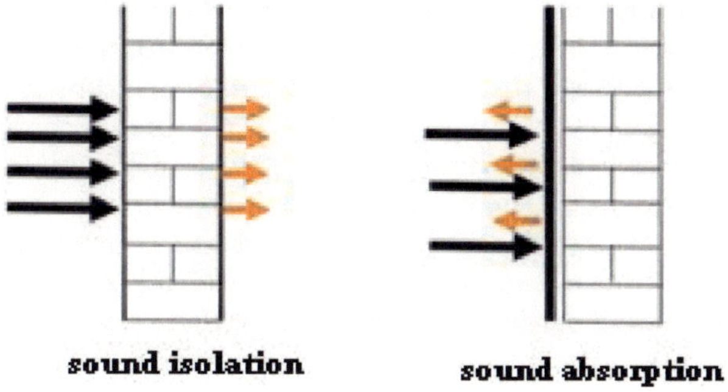

Figure 1.
Sound isolation and absorption [23].

source. Since frictional and momentum losses and heat fluctuation occur as the sound goes through the sound absorbing material, these are the essential reasons of the acoustic energy loss [2, 10].

Sound insulation deals with the transmission of sound between the adjacent rooms. Sound transmission loss indicates the sound insulation ability of a material and is denoted as the difference between the transmitted sound power and the sound power level of the incident wave [24]. An applying barrier to prevent transmission is the main principle. Materials that have high density such as thick glass, solid, metal, and brick are used for this aim [5].

Sound absorptive materials are known as the materials reducing the acoustic energy of a sound wave as the wave passes through it. They are usually used to soften the acoustic environment of a closed volume by reducing the amplitude of the reflected waves [25]. There are three types of absorbers: porous absorbers, membrane absorbers, and resonance absorbers. Inside the porous absorbers, sound transmission occurs in such a way that viscous and thermal effects cause acoustic energy to be dissipated. They are usually fuzzy, fibrous materials such as textiles, mineral wool, curtains, clothing, carpets, and certain types of foam plastic. As the sound energy penetrates into the material on hitting the surface, the sound-absorbing effect is obtained. Generally, since the required thickness is large with porous textile absorbers, obtaining adequate absorption at low frequencies is difficult. Membrane absorber is basically a flat box, 100–200 mm deep, installed in the wall with a thin sheet of plywood or similar on the front and with a light mineral wool filling the box cavity. Resonance absorber takes the form of perforated plasterboard, perforated metal corrugated sheets, and metal boxes. The working principle of the resonant absorbers is dissipating acoustic energy with structure vibration. Resonance absorbers are usually efficient in a thin tunable low-frequency band, whereas the porous absorbers are good for mid-to-high frequency range [26, 27].

Even though all materials absorb a part of incident sound, acoustic material term is used, especially for materials providing high values of absorption for the specific application fields. The acoustic materials are used in various applications such as in building construction; transportation; industries for specific purposes; such as concert rooms, schools, theaters, recording studios, lecture halls, and so forth. In recent days, acoustic textiles as acoustic materials are critical for the reduction of noise pollution.

The sound intensity, sound pressure levels, and sound classification systems can be used to assess the performance of the acoustic structure. Sound absorption defines the energy amount absorbed by the material and stated as sound absorption coefficient (α) ranging between 0 and 1, where 0 means no absorption and 1 means the highest or total absorption. The higher coefficient expresses lower reverberation time. The reverberation time is the time lag, in seconds, measured for the sound to decay by 60 dB in a space after a sound source has been stopped. The noise reduction coefficient is the average absorption coefficients of an acoustic material at a typical frequency set of 250, 512, 1024, and 2048 Hz defined according to the tube type and acoustic measuring instrument used for the tests [5, 24].

Nonwoven materials are ideal materials for sound insulation and sound absorption applications in order to decrease sound pollution in the environment due to their fibrous structure and high total surface area [28]. Areal density (mass), porosity, volumetric density, tortuosity, particle size distribution, and thickness constitute significant physical properties of nonwoven fabrics for acoustic applications. Acoustic ceilings, noise reducing quilts, and noise proof barriers are some of the applications of nonwoven fabrics serving as noise absorption elements. A wide variety of studies on the acoustic properties of nonwoven products are available, which some of them are given in the following paragraphs.

Lee and Joo investigated the usage of the recycled polyester nonwovens as a sound absorber instead of conventional materials such as glass wool and rockwool by using a two-microphone impedance tube. Nonwoven having more fine fiber was found to be better at contacting the sound wave due to more resistant characteristic. The nonwoven absorber having an unoriented web in the middle layer had a higher noise absorption coefficient (NAC) than the ones having entirely oriented web structure, but the difference was insignificant. The panel resonance effect had contributed to increase the noise absorption coefficient. In the case of coating structure, the panel promoted the NAC in low- and middle-frequency regions, but reverse effect was obtained in the high frequency region by the coincidence effect [22].

Na et al. measured the sound absorption features of the fabric-nonwoven system produced by adding a microfiber fabric to a layer of 15-, 30-, and 45-mm nonwoven by reverberation chamber method. The results revealed that the fabrics made from microfibers were quite advantageous compared to the conventional fabrics of the similar thickness or weight in terms of sound absorption characteristics [29].

Tascan and Vaughn studied the effects of total fiber surface area and fabric density on needle punched nonwovens. It was reported that the needle punched nonwoven fabrics produced from polyester fibers having octalobal and trilobal cross-sectional shape had better sound insulation results than the nonwoven fabrics produced from round fibers. Moreover, nonwovens with finer fibers in various cross-sectional shapes had better sound absorption and insulation than the ones made from coarser fibers. Fabric density and total fiber surface area in needle punched nonwoven fabrics were found to be in the tendency of improving fabric sound insulation [28].

Sengupta searched the effect of fabric type, density, the number of layers, source intensity, the distance of fabric from the receiver, the distance of the fabric from the sound source, and fiber type on the sound reduction of various needle-punched nonwoven fabrics. A sound insulation box was used for measurement. It was found that higher area density was one of the reasons of higher sound reduction. A negative correlation between the area density and bulk density of needle punched nonwoven and sound reduction was determined. Moreover, maximum sound reduction among jute, polypropylene, polyester, and other jute-polypropylene blended (3:1 and 1:3) nonwovens was obtained by jute-polypropylene (1:1) blend [30, 31].

Küçük and Korkmaz tested eight different nonwoven composites produced from various blends. Sound absorption properties, weight per unit area, thickness, and air permeability parameters of the samples were measured. An increase in sound absorption properties of the material was determined along with the increase in thickness and the decrease in air permeability. About 70% cotton and 30%

polyester nonwoven supplied the best sound absorption coefficient in the mid-to-high frequency ranges. It was revealed that the increase in the amount of fiber per unit area caused an increase in sound absorption of the material. The results indicated that acrylic and polypropylene addition into a cotton and polyester fiber mixture increased the sound absorption properties of the composite in the low- and mid-frequency ranges [3].

Recently, recycled nonwovens have become important sound absorption materials. Some of the studies related to sound absorption characteristic of the recycled nonwoven surfaces are given in the following paragraphs.

Seddeq et al. searched the acoustic properties of the nonwovens produced from recycled natural fibers blended with synthetic fibers and studied PET/cotton-wool (70/30), PP/cotton-wool (80/20), cotton/PET (50/50) blends and 100% jute fibers in the range of 422–561 g/m^2 and 2.53–5.64 mm. At high frequencies (2000–6300 Hz), nonwoven samples had high sound absorption coefficients, whereas at low frequencies (100–400 Hz), low sound absorption coefficients were obtained, and at mid frequencies (500–1600 Hz), better sound absorption coefficients were found. The increased thickness of nonwovens improved the sound absorption coefficients of the materials at all frequency ranges [10].

Carvalho et al. studied qualitative analysis of the acoustic insulation behavior of various thermo-bonded nonwoven fabrics. Nonwoven fabrics were produced from mineral wool and recycled fibers, containing a mixture of polypropylene, cotton, acrylic, and polyester. Some samples were laminated with aluminum foil. Thermo-bonded nonwovens with high thickness value and laminated with aluminum foil displayed better sound reduction performance than the other single-layered nonwovens made from recycled fibers. They showed better performance than the nonwovens made from mineral wool as well [32].

Manning and Panneton studied the acoustic properties of post-consumer and industrial recycled fiber absorbers. Three different shoddy samples, needle-punched mat, thermally bonded mat, and resin-bonded mat, were compared in the study. A lower noise absorption coefficient value of 0.20 was obtained at low frequency (0–1000 Hz) for all samples. The results showed that the noise absorption coefficient values of the samples in frequency range of 0–4000 Hz were similar to each other [33].

Rey et al. designed novel green sound absorbing materials as a part of noise barriers. They used recycled textile materials and nontoxic binder fibers to manufacture the eco-materials. Acoustic characterization of noise barriers was measured in a small-scale reverberation room designed for the testing of small samples. New materials used in noise barrier prototypes performed very well in accordance with the international standards. The performance of them was comparable with those of commercially available noise barriers made of typical sound absorbing materials [34].

Patnaik et al. studied the sound insulation properties of needle-punched samples produced from waste wool and recycled polyester fibers (r-PET) for building industrial applications. Waste wool fibers were mixed with r-PET fibers in 50/50 proportion to prepare needle-punched mats, and their acoustic properties were evaluated with other performance properties. Nonwoven mats were produced from waste wool fibers that are coring wool (CW) and Dorper wool (DW); r-PET fibers; and blended r-PET fibers in 50/50 proportion with these fibers (DWP, CWP). Good sound absorption properties in the overall frequency range (50–5700 Hz) were obtained for all nonwoven samples. The sound absorption was lower at low frequencies (50–1000 Hz) and increased from medium (1000–2000 Hz) to high frequency range (200–5700 Hz) for all the samples. The lowest noise absorption coefficient value was 0.61 for r-PET, and the highest was 0.75 for DWP for all frequency ranges.

DWP presented higher α value than CWP due to the presence of longer fiber length. It was stated that sound absorption depends on thickness of the material among other factors. They also determined that the r-PET/wool mats (DWP, CWP) could absorb more than 70% of the incident noise in the overall frequency range [35].

Kalebek investigated acoustic behavior of needle-punched nonwoven fabrics produced from recycled PES fibers for the automotive industry. The physical properties such as density, thickness, weight per unit area, air permeability, tensile strength, and elongation were measured and compared to each other. The fabric mass per unit area and the thickness of the needle-punched nonwoven fabrics were found to be positively effective on the sound insulation. Additionally, it was observed that higher air permeability caused higher sound transmission and, as a result, lower sound insulation [30].

3.2 Measurement methods of acoustic absorption properties of materials

There are various standards related to test procedures for determination of acoustic features of textile materials. In terms of standards, sound absorption properties of nonwovens can be defined in different parameters such as sound absorption coefficient, transmission coefficient, reflection coefficient, sound transmission loss, airflow resistivity, and sound power ratio. Some of the commonly used standards are as follows: ISO 354:07 Acoustics—measurement of sound absorption in a reverberation room; ISO 11957:2009 Acoustics—determination of sound insulation performance of cabins (laboratory and in situ measurements); ISO 10534-1:96 Acoustics—determination of sound absorption coefficient and impedance in impedance tubes—Part 1: method using standing wave ratio; ISO 10534-2:98 Acoustics—determination of sound absorption coefficient and impedance in an impedance tube—Part 2: transfer-function method; ASTM E2611-19 Standard test method for measurement of normal incidence sound transmission of acoustic material based on the transfer matrix method; ASTM E1050-19 Standard test method for impedance and absorption of acoustic materials using a tube, two microphones, and digital frequency analysis system; and ASTM C423-17 Standard test method for sound absorption and sound absorption coefficients by the reverberation room method.

The impedance tube method and alpha cabinet methods, which are used in experimental part, will be explained in detail in the following paragraphs.

3.2.1 Measurement of acoustic absorption characteristics by impedance tube method

One of the methods mostly utilized to determine the sound absorption coefficients of textile surfaces is the ISO 10534-2 double microphone impedance tube method. The sound pressure difference with the help of a microphone placed in the impedance tube is measured. The sound is created by the signal generator in the apparatus to define the sound absorption coefficient and transmitted through the impedance tube. The performance of the material is examined by the software.

The measurement principle of the impedance tube method depends on measuring the reflected sound wave and calculating the sound absorption coefficient from the surface impedance value and transfer function. Obtaining the surface impedance and sound absorption coefficient values in one measurement for each frequency generates the advantage of this method [36].

During the measurement, sound is generated by the sound source, and the receiving decibels are evaluated by the decibel meter with and without the sample, and subsequently, the sound insulation by the fabric samples is calculated [37]. The tube is produced by rigid, transparent, or opaque materials to limit the sound within

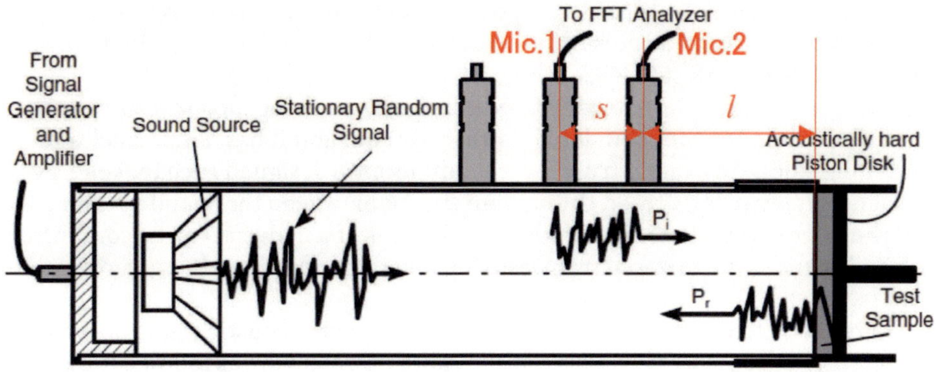

Figure 2.
Impedance tube sound absorption measurement method [41].

the tube along one direction toward the direction of transmission. Consequently, it simplifies the three-dimensional wave equations to one-dimensional wave equation. The sample size is small compared to reverberation method [38].

In the impedance tube arrangement schematically shown in **Figure 2**, first, the test sample is attached at one end of the impedance tube, and the signal characterizing the voice is generated by the software of test system. Plane waves are created in the tube by sound source emitting random or pseudo-random cycle, and pressure is measured at two places close to the sample. Then, the signal passing across the amplifier is transformed into a planar progressive sound wave in the tube through the speaker. The transfer function between two microphones is a ratio of the pressure values measured separately from the two microphones. The transfer function is related to the value of the reflection factor and the value of the sound absorption coefficient of that frequency (R) is obtained from this factor [36]. The frequency scale is based on the diameter of the tube and the gap between the two microphones. With the impedance tube method, the sound absorption coefficient values of the materials can be measured in the frequency range of 50 Hz to 6.4 kHz.

The normal incidence reflection factor is computed by Eq. (1).

$$r = |r| e^{j\phi r} = \frac{H_{12} - H_1}{H_R - H_{12}} e^{2jk_0 x_1} \tag{1}$$

where r is a reflection factor of normal incidence; x_1 is the distance between the sample and the further microphone location; j is the square root of minus one; k is $2\,\pi\,f/c$ (m^{-1}); Φr is the phase angle of the normal incidence reflection factor; H_{12} is the transfer function from microphone one to two, defined by the complex ratio $p2/p1 = S_{12}/S_{21}$; and H_R and H_I are the real and imaginary part of $H_{12.}$

The sound absorption coefficient (α) is calculated by Eq. (2) [39, 40]:

$$\alpha = 1 - |r|^2. \tag{2}$$

3.2.2 Measurement of acoustic absorption characteristics by alpha cabinet method

ISO 354 and ASTM C423-17 standards explain the measurement method of sound absorption by the reverberation room. In reverberation room, reverberant sound field closely approaches a diffuse sound field. This approach occurs, when the source is on, that is the stable condition and also after the sound source is stopped, which is defined as decomposition condition. The room is isolated adequately to keep outside noises and structural vibrations from impeding during the measurements (ASTM C423-17).

Reverberation room is naturally large empty room having long reverberation times. The volume of the room is higher than 200 m^3, and it has nonparallel wall and ceiling sides. An exactly truly reverberant room is designed such that energy from a noise source is diffused throughout the room to keep the sound pressure level the same everywhere. Absorption coefficients of a material can be computed with and without the material by measuring reverberation times in a reverberation room [42].

There are different types of cabinets used to measure acoustic sound absorption characteristics on the market. Some of the cabin examples are as follows: Rieter Alpha Cabin Instructions: Technical Note 591; Toyota Engineering Standard: test method for acoustic materials, TSL0600G; Renault Test Method: fibrous and cellular material sound absorption in diffuser field, D49-1977-B; and the design of small reverberation chambers for transmission loss measurements [36].

The name of alpha cabinet comes from the name of the sound absorption coefficient "alpha," and their structure is like a miniature reverberation room (**Figure 3**). During the measurements, the relative humidity in the alpha cabinet is 55%, and the temperature is set at 25°C. Before testing in the alpha cabinet, it is necessary to compare the noise levels in the cabinet with the background noise levels from outside (motor noise and electrical noise), while the speakers are in operation. The difference between these levels should be at least 45 dB. Alpha cabinets are quite similar to the reverberation rooms (**Figure 4**).

Sabine's reverberation formula is used for determining the material absorption coefficients with the help of reverberation rooms and is still commonly accepted as a very practical evaluation method for the reverberation time in rooms. Sabin is a unit of sound absorption in square meter, indicating the area of open window. The main principle of the method is that sound energy moving in the direction of an open window in a room will not be reflected at all; however, it entirely disappears in the open air outside. The same effect will be obtained if the open window would be replaced with 100% absorbing material in the same sizes [45].

The Sabine formula is applied for the sound absorption coefficient measurements, and the coefficient is analyzed by taking the difference between the resonance times of the sampled and nonsampled measurements in the cabinet [36].

$$\alpha = \frac{0.163 \times V}{S}\left(\frac{1}{TR} - \frac{1}{TR_0}\right) \times C \tag{3}$$

where V is the cabinet volume, S is the sample area, TR is the sound receiving time with the test sample inside the cabinet, TR_0 is the sound receiving time without the test sample inside the cabinet, and C is the cabinet correction coefficient.

Sabine also derived a definition for the time, T, of the residual sound to decay below the audible intensity, starting from 1,000,000 times higher initial amount given in Eq. (4).

$$T = 0.161\ V/A \tag{4}$$

where V is the room volume in m^3 and A is the total absorption area in m^2 [45, 46].

The noise reduction coefficient (NRC) is calculated by deriving the arithmetic mean of the absorption coefficients in the 250, 500, 1000, and 2000 Hz 1/3-octave frequency bands. This number is rounded to the nearest multiple of 0.05. The zero

Figure 3.
Twin mini reverberation cabinets [43].

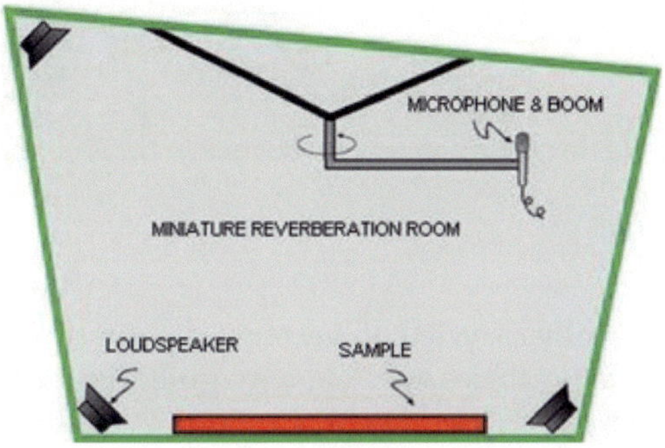

Figure 4.
Alpha cabinet measurement arrangement [44].

value of NRC indicates the perfect reflection. The NRC value 1 means that the perfect absorption happens. Reverberation room techniques can also be used to measure this parameter [5].

When studies carried out in the literature are examined, it can be concluded that impedance tube measurements and alpha cabinet measurements do not give comparable results, as also found in this study. Therefore, correlation analysis between the results is not necessary.

4. Experimental part

Today, while studies on new sound insulating materials continue intensively, the usage of recycled products in this area is also increasing. Recycling technology provides environmental, social, and economic benefits. With environmental awareness and increased sensitivity to sustainability, recycled materials provide significant advantages for sound insulation.

In this study, nonwoven materials produced from conventional polyester (PES), mechanically recycled polyester (rm-PES), recycled polyester (r-PET), polypropylene (PP), and recycled polypropylene (r-PP) were compared with each other in terms of their sound absorption characteristics [47]. In **Figure 5**, photographs of the samples are given.

Supplying the raw materials and producing the nonwoven fabrics were carried out in a Turkish company working especially on recycled products. The processes

Figure 5.
Photographs of the samples used in experimental.

applied for the mechanically recycled polyester and polypropylene fibers were as follows: the collected waste fabrics were cut with guillotine cutting machine and then shredded into pieces with particles of a suitable size. Then, the pieces were weighted on an electronic balance and blended in a vertical mixing blender. Carding process was applied to break down the big fiber bundle to the small size. Two-time carding process was applied to get uniform web formation, and then, the fine opener was used.

r-PET fibers provided by using r-PET flakes gained from PET bottles were used in this study. The common method of fiber production is as follows: at first, PET bottle wastes are separated from the other wastes; then, they are broken into flakes, washed, and dried before the spinning process of the fiber. In this method, r-PET fibers are acquired by transforming PET chips by the melt fiber drawing method [17].

The nonwoven fabric samples were produced by needle-punching technology. The webs were fed to a needling loom to produce nonwoven surfaces. Softness, bulkiness, conformability, fibrousness, and high strength characteristic without binder usage make the needle-punched nonwoven fabrics unique among the other nonwovens.

Before the physical tests, nonwoven samples were conditioned in standard atmospheric conditions (20 ± 2°C temperature, 65 ± 4% relative humidity). The mass per unit area was determined in accordance with the ISO 9073-1 standard, and thickness measurements of the materials were carried out according to the ISO 9073-2 standard using SDL Atlas fabric thickness gauge. Air permeability test was conducted according to the EN ISO 9237 standard by using Textest FX 3300 air permeability meter at 200 Pa pressure difference and 20 cm^2 measurement areas.

Two methods were applied to measure sound absorption properties of the samples: alpha cabinet and impedance tube methods. Nonwoven textile products supplied within the scope of the study were tested by using alpha cabinet according to the Renault D49-1977-B standard. The cabinet volume was 6.44 m^3, and the total surface area was 22.2 m^2. The surface area of the test specimens was 1.12 m^2; the height of the microphone from the ground was 0.88 m; and the height of the sound source was 0.20 m. The frequency measurement range was set at 400–10,000 Hz. Three Brüel and Kjaer 2669 model microphones were used.

Impedance tube method was also used to measure the sound absorption coefficients of the samples according to the EN ISO 10534-2 method. The frequency was set from 100 to 1600 Hz. The test was carried out by three test samples randomly taken from nonwoven fabrics.

The structural properties and air permeability values of five different nonwoven surfaces used in the study are given in **Table 1**.

As air permeability results of nonwoven fabrics are examined, the surface produced from r-PP material has the highest air permeability, whereas the lowest air permeability value belongs to mechanically recycled polyester surface.

The test results obtained from the impedance tube method are given in **Figure 6**. According to the results, at low frequencies between 100 and 400 Hz, it can be clearly seen that r-PP, rm-PES, and r-PET surfaces have higher sound absorption coefficient values than the conventional PES and PP fabrics. This is an important point that recycled textile surfaces can be suggestible as sound insulation materials in low frequency band gap (100–400 Hz) as an alternative to the conventional fibers. In addition to this, r-PP material has the highest sound absorption coefficient (over 0.50) among the other recycled products since its lower density and porous structure create a higher friction surface. As the performance at the mid frequencies (400–1600 Hz) is analyzed, only PP, r-PP, and rm-PES fabrics have sound absorption coefficient over 0.50. Therefore, these surfaces can be suggested for sound insulation materials to be used at mid frequencies.

Compared to conventional polypropylene, r-PP indicates better sound absorption characteristics for all frequencies. It is associated with the micro voids in the recycled material caused by the inhomogeneity [48].

The acoustic absorption test results of the textile surfaces according to the alpha cabinet method are given in **Figure 7**.

Measured parameter		**100% PES**	**100% r-PET**	**100% rm-PES**	**100% PP**	**100% r-PP**
Mass per unit area (g/m^2)	Mean ($\bar{X}$)	550	596	519	509	500
	Standard deviation (SD)	8.14	12.10	27.27	10.92	4.81
	CV (%)	1.48	2.03	5.25	2.15	0.96
Thickness (mm)	Mean ($\bar{X}$)	4.04	2.95	3.17	3.60	4.55
	Standard deviation (SD)	0.16	0.09	0.30	0.10	0.10
	CV (%)	3.93	2.89	9.46	2.67	2.28
Air permeability (l/m^2 s)	Mean ($\bar{X}$)	1136	1270	600	875	1536
	Standard deviation (SD)	57.71	116.40	112.13	148.66	69.14
	CV (%)	5.08	9.17	18.66	17.01	4.50
Fabric density (g/cm^3)	Mean ($\bar{X}$)	0.136	0.202	0.164	0.141	0.110

Table 1.
Properties of nonwoven fabrics used in the study.

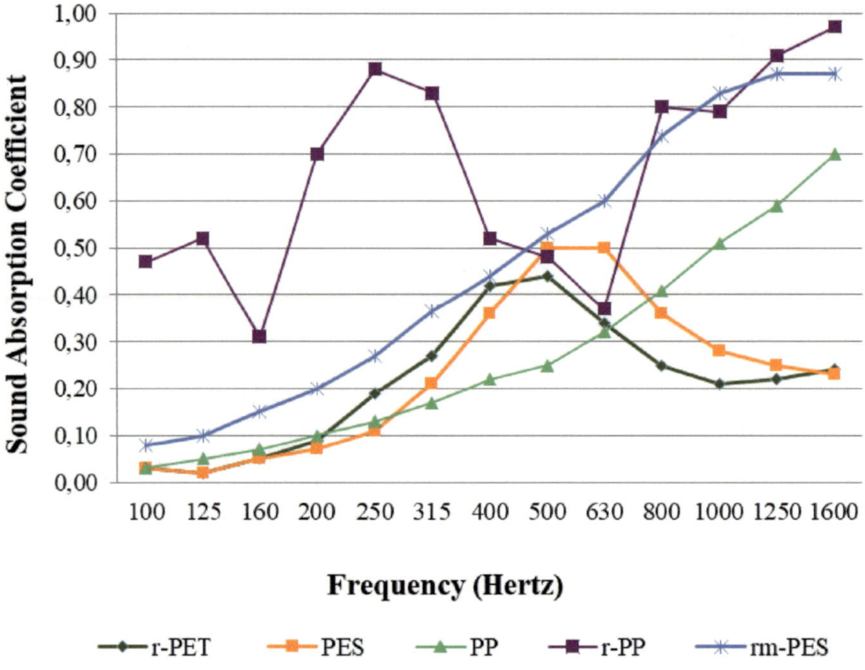

Figure 6.
Sound absorption coefficients based on the impedance tube method.

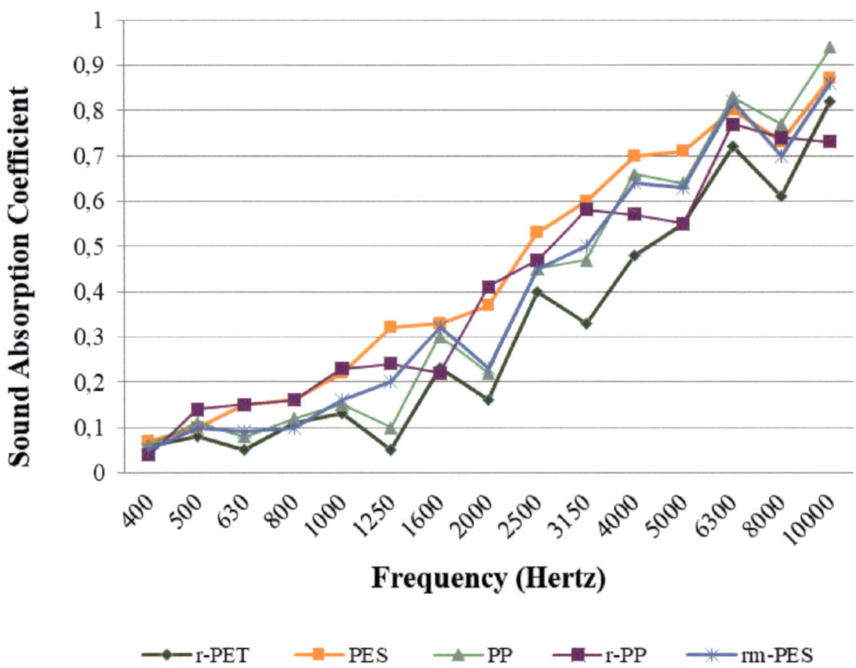

Figure 7.
Sound absorption coefficients based on alpha cabinet method.

Materials can be tested at a wide frequency band in case of their sound absorption properties. In this study, the tested frequency range is 400–10,000 Hz. According to **Figure 7**, between 400 and 1600 Hz, sound absorption coefficients of all nonwoven surfaces are close to each other and are lower than 0.35. Over 2500 Hz frequency, which is a high frequency level, sound absorption coefficients of the nonwovens are over 0.5, which supply a better sound insulation property and can be an alternating insulation material for the industry.

In this study, acoustic absorption coefficients were measured according to two different sound measurement methods. The results obtained from the alpha cabinet and the impedance tube methods complement each other and give the idea for the performance and comparability of the materials at different sound frequencies. For obtaining the required results, the choice of the appropriate method is crucial. The impedance tube method is mostly suitable in order to develop new materials, while a reverberation room test is more relevant to design acoustic adjustments of the space [49]. Moreover, reverberation rooms are often preferred since they give closer results to the actual working conditions of the material by means of the random incidence excitation. However, the tests in impedance tube allow obtaining the additional information about the characteristic acoustic impedance of the sample [50].

5. Conclusion

Along with the growing population, noise has become one of the major problems of everyday life, affecting our quality of lives and, in some cases, our health. As the studies on reducing the noise generation continue, new solutions are being searched on systems that allow absorbing more quantities of the present disturbing noise. Within this scope, many different technical textile materials are produced, which are closely related to building construction, automotive, and machinery industries. The use of these high-performance products is becoming increasingly widespread in terms of technical specifications. However, in today's competitive conditions, products that can compete in terms of cost are more preferred in the market. Recycled surfaces are one of these preferred products. In addition to the price advantage, waste materials are transformed into usable products by recycling, which is completely an environmentally friendly production that provides sustainability.

In this study, characteristics of conventional polyester and polypropylene nonwoven insulation materials commonly used in the market and the nonwoven fabrics produced from recycled materials were compared. When the test results were examined, it was concluded that the recycled materials used in sound insulation area had very successful competitive performance when compared with the conventional materials. They had the qualifications supplying the expectations in terms of sound insulation when produced in enough thickness value.

Insulation materials are used, especially in areas where sound insulation is desired (e.g., children's houses, hospitals, entertainment places, and automotive sector) and in civil engineering area. In addition, it is a good application area for recycled materials to supply a sustainable world, to increase the environment friendly approach, and to evaluate the waste materials in a technical field. By the year 2019, price of PP fiber in the market is about 3.2 $/kg, 1.6 $/kg for r-PP fiber, about 1.6 $/kg for 100% PES and r-PET fibers, and 1.2 $/kg for rm-PES fibers. In the scope of economy, it can be foreseen that recycling materials will provide an important advantage for both producers and consumers.

Acknowledgements

This study was supported by the Ege University Scientific Research Projects Coordination Unit (Project Number: 15-MÜH-078). The authors would also like to thank Sarıkılıç Tekstil San ve Tic. Ltd. Şti. for their material support used in the study. The experimental part of this chapter was presented in 1st International Technology Sciences and Design Symposium ITESDES Giresun, Türkiye, June 28–30, 2018.

Author details

Nilgün Özdil[1*], Gonca Özçelik Kayseri[2] and Gamze Süpüren Mengüç[2]

1 Ege University, Department of Textile Engineering, İzmir, Turkey

2 Ege University, Emel Akın Vocational Training School, İzmir, Turkey

*Address all correspondence to: nilgun.ozdil@ege.edu.tr

References

[1] Wintzel L. Acoustic textiles: The case of wall panels for home environment [Thesis]. Borås, Sweden: The Swedish School of Textiles; 2014

[2] Kaya Aİ, Dalgar T. Ses yalitimi açisindan doğal liflerin akustik Özellikleri, Makü Febed. Özel Sayı. 2017;**1**:25-37

[3] Küçük M, Korkmaz Y. The effect of physical parameters on sound absorption properties of natural fiber mixed nonwoven composites. Textile Research Journal. 2012;**82**(20):2043-2053. DOI: 10.1177/0040517512441987

[4] Kadam V, Nayak R. Basics of acoustic science. In: Padhe R, Nayak R, editors. Acoustic Textiles, Textile Science and Clothing Technology. Singapore: Springer; 2016. DOI: 10.1007/978-981-10-1476-5. ISBN: 978-981-10-1474-1

[5] Padhye R, Nayak R. Acoustic textiles: An introduction. In: Padhe R, Nayak R, editors. Acoustic Textiles, Textile Science and Clothing Technology. Singapore: Springer; 2016. DOI: 10.1007/978-981-10-1476-5. ISBN: 978-981-10-1474-1

[6] Harris DA. Sound absorbing materials. In: Noise Control Manual. Boston: Springer; 1991. DOI: 10.1007/978-1-4757-6009-5_2

[7] Arenas J, Applications of acoustic textiles in automotive/transportation. In: Padhe R, Nayak R, editors. Acoustic Textiles, Textile Science and Clothing Technology. Singapore: Springer; 2016. DOI: 10.1007/978-981-10-1476-5

[8] Stanciu MD, Curtu I, Cosereanu C, Lica D, Nastac S. Research regarding acoustical properties of recycled composites. In: 8th International DAAAM Baltic Conference, Industrial Engineering, Tallinn, Estonia; 2012

[9] Trajković D, Jordeva S, Tomovska E, Zafirova K. Polyester apparel cutting waste as insulation material. The Journal of the Textile Institute. 2016;**108**(7):1238-1245. DOI: 10.1080/00405000.2016.1237335

[10] Seddeq SH, Aly NM, Marwa A, Elshakankery MH. Investigation on sound absorption properties for recycled fibrous materials. Journal of Industrial Textiles. 2012;**43**(1):56-73. DOI: 10.1177/1528083712446956

[11] Technical Insulation. Available from: https://www.sandler.de/en/sawatec-sound-and-heat-insulation [Accessed: 06 April 2020]

[12] Chavan RB. Environmental sustainability through textile recycling. Journal of Textile Science & Engineering. 2014:1-5. DOI: 10.4172/2165-8064.S2-007. ISSN: 2165-8064

[13] Eser B, Çelik P, Çay A, Akgümüş D. Tekstil ve konfeksiyon sektöründe sürdürülebilirlik ve geri dönüşüm olanakları. Tekstil ve Mühendis. 2016;**23**(101):43-60. DOI: 10.7216/1300759920162310105

[14] Bartl A. Moving from recycling to waste prevention: A review of barriers and enables. Waste Management & Research. 2014;**32**:3-18

[15] Piribauer B, Bartl A. Textile recycling processes, state of the art and current developments: A mini review. Waste Management & Research. 2019;**37**(2):112-119. DOI: 10.1177/0734242X18819277

[16] Hole G, Hole AS. Recycling as the way to greener production: A mini review. Journal of Cleaner Production. 2019;**212**:910-915. DOI: 10.1016/j.jclepro.2018.12.080

[17] Telli A, Özdil N. Properties of the yarns produced from r-PET fibers and their blends. Tekstil ve Konfeksiyon. 2013;**23**(1):3-10

[18] Hawley JM. Textile recyling: A system perspective. In: Wang Y, editor. Recycling in Textiles, 1st Edition. Cambridge: Textile Institute, Woodhead Publishing; 2006. DOI: 10.1533/9781845691424.1.7

[19] Borman T, Sun D. Recycled jean: Property comparison to standard Jean. Journal of Fashion Technology and Textile Engineering. 2016;**4**(2):1-10. DOI: 10.4172/2329-9568.1000136

[20] Alan G. Tıraşlama artiği kirpinti liflerin değerlendirilmesi üzerine bir araştirma [PhD thesis]. Uşak: Uşak Üniversitesi Fen Bilimleri Enstitüsü; 2018

[21] Santhanam S, Bharani M, Temesgen S, Atalie D, Ashagre G. Recycling of cotton and polyester fibers to produce nonwoven fabric for functional sound absorption material. Journal of Natural Fibers. 2018;**16**(2):300-306. DOI: 10.1080/15440478.2017.1418472

[22] Lee Y, Joo C. Sound absorption properties of recycled polyester fibrous assembly absorbers. AUTEX Research Journal. 2003;**3**(2):78-84

[23] Kaya Aİ. Kompozit malzemeler ve özellikleri. Putech & Composite Poliüretan ve Kompozit Sanayi Dergisi. 2016;**29**:38-45

[24] Zhu X, Kim B, Wang Q, Wu Q. Recent advances in the sound insulation properties of bio-based materials. BioResources. 2014;**9**(1):1764-1786

[25] Bajaj S. Electronics and computing in textiles—A breakthrough innovation. In: National Conference on Recent Trends in Environment. Science and Technology; 2016. pp. 68-71. ISBN: 978-93-84935-82-5

[26] Absorber types within room acoustics. Available from: https://www.troldtekt.com/Product-properties/Good acoustics/Advanced_acoustics/Different-absorber-types [Accessed: 10 April 2020]

[27] Xiaojun Q. Principles of sound absorbers. In: Padhye R, Nayak R, editors. Acoustic Textiles, Textile Science and Clothing Technology. Singapore: Springer; 2016. pp. 43-72. DOI: 10.1007/978-981-10-1476-5. ISBN: 978-981-10-1476-5

[28] Tascan M, Vaughn E. Effects of total surface area and fabric density on the acoustical behavior of needle-punched nonwoven fabrics. Textile Research Journal. 2008;**78**:289-296. DOI: 10.1177/0040517507084283

[29] Na Y, Lancaster J, Casali J, Cho G. Sound absorption coefficients of micro-fiber fabrics by reverberation room method. Textile Research Journal. 2007;**77**(5):330-335. DOI: 10.1177/0040517507078743

[30] Kalebek AN. Sound absorbing polyester recycled nonwovens for the automotive industry. Fibres & Textiles in Eastern Europe. 2016;**24**(115):107-113. DOI: 10.5604/12303666.1172093

[31] Sengupta S. Sound reduction by needle-punched nonwoven fabrics. Indian Journal of Fibre & Textile Research. 2010;**35**:237-242

[32] Carvalho R, Rana S, Fangueiro R, Soutinho F. Noise reduction performance of thermobonded nonwovens. In: 12th World Textile Conference AUTEX; Zadar, Croatia; 13-15 June, 2012

[33] Manning J, Panneton R. Acoustical model for shoddy-based fibre sound absorbers. Textile Research Journal. 2013;**83**(13):1356-1370. DOI: 10.1177/0040517512470196

[34] Rey R, Bertób L, Albac J, Arenasd JP. Acoustic characterization of recycled textile materials used as core elements in noise barriers. Noise Control Engineering Journal. 2015;**63**(5):439-447. DOI: 10.3397/1/376339

[35] Patnaik A, Mvubu M, Muniyasamy S, Botha A, Anandjiwala RD. Thermal and sound insulation materials from waste wool and recycled polyester fibres and their biodegradation studies. Energy and Buildings. 2015;**92**(1):161-169. DOI: 10.1016/j.enbuild.2015.01.056

[36] Uzundağ U, Tandoğan O. Malzemelerin Akustik Performans Testleri. Türkiye: Novosim Mühendislik Hizmetleri; 2013

[37] Fera TF. Manufacture of reclaimed fiber non-woven for sound absorption. Journal of Materials Science and Engineering. 2018;7:503. DOI: 10.4172/2169-0022.1000503

[38] Niresh J, Neelakrishnan S, Subharani S, Prabhakaran R. Performance testing for sound absorption coefficient by using impedance tube, research. Journal of Applied Sciences, Engineering and Technology. 2015;**11**(2):185-189. DOI: 10.19026/rjaset.11.1706. ISSN: 2040-7459; e-ISSN: 2040-7467

[39] Bhattacharya SS, Bihola DV. Development of impedance tube to measure sound absorption coefficient. International Journal of Engineering and Advanced Technology (IJEAT). 2019;**8**(6):3218-3222. ISSN: 2249-8958

[40] ISO 10534-2:1998. Acoustics-Determination of Sound Absorption Coefficient in Impedance Tubes—Part 2: Transfer-Function Method

[41] Kimura M, Kunio J, Schuhmacher A, Ryu Y. A new high-frequency impedance tube for measuring sound absorption coefficient and sound transmission loss. In: Inter-Noise Conference; 2014

[42] Parkinson JP. Acoustic absorber design [Master of Engineering thesis]. Christchurch, New Zealand: University of Canterbury Department of Mechanical Engineering; 1999

[43] Akustik Test Odaları ve Kabinler, Available from: http://proplan.com.tr/Uzmanliklar/endustriyel-ve-mimari-akustik/test-odalari-ve-kabinler [Accessed: 10 April 2020]

[44] Acoustic Materials Development Noise and Vibration Engineering. Available from: https://www.roush.com/downloads/acoustic-material-development/?wpdmdl=3521 [Accessed: 10 April 2020]

[45] Taşcan M. Acoustical test methods for nonwoven fabrics. In: Padhye R, Nayak R, editors. Acoustic Textiles, Textile Science and Clothing Technology. Singapore: Springer; 2016. DOI: 10.1007/978-981-10-1476-5. ISBN: 978-981-10-1476-5

[46] Atmajan A, Tessy I, Abin M, Pisharady SK. Chapter 7: Acoustics. In: A Textbook of Engineering Physics. New Delhi: Acme Learning Private Limited; 2011. p. 107. ISBN: 8121908175, 9788121908177

[47] Kayseri GÖ, Mengüç GS, Özdil N. Sound absorption characteristics of textile materials produced from recycled fibers. In: 1st International Technology Sciences and Design Symposium; Giresun; 2018. pp. 1372-1383

[48] Murugan D, Varughese S, Swaminathan T. Recycled polyolefin based plastic wastes for sound absorption. Polymer - Plastics Technology and Engineering. 2006;**45**(7):885-888. DOI: 10.1080/03602550600611818

[49] Drabek P. Comparison of two internationally recognized methods for determining the sound absorption coefficient. In: MATEC Web of Conferences (CSCC 2017). Vol. 125; 2017. p. 02030

[50] Piana EA, Roozen NB, Scrosati C. Impedance tube measurements on the denorms round Robin test material samples. In: 26th International Congress on Sound and Vibration; Montreal; 7-11 July 2019

Chapter 6

Innovation of Textiles through Natural By-Products and Wastes

Lorena Coelho, Ana Isabel Magalhães, Sara Fernandes, Patrícia Batista, Manuela Pintado, Pedro Faria, Catarina Costa, Bruna Moura, Augusta Marinho, Rosa Maria, Albertina Reis, Marta Carvalho, Mário Marques, Ângela Teles, José De Almeida Morgado, Maria Helena Vilaça, Jéssica Alexandra Pereira, Pedro José Magalhães, Ana Sofia Silva, Ricardo Jorge Silva, Mário Jorge Silva, Vera Lúcia Sá, Sandra Gabriela Ventura, João Silva Abreu, Joaquim Manuel Gaião, Raquel Rosa Mourão, Fernando Manuel Merino, Mónica Sofia Gonçalves and Regina Malgueiro

Abstract

Nowadays, the competitiveness of the textile industry and the consumers' interest have been increasing the demand for innovative and functional textiles. Allied to this, sustainable developments are playing an increasingly important role in the textile industry. Such concerns led to a new development strategy based on the valorization of bio-based wastes and by-products of different industries, inserting this in the circular economy paradigm. These bio-based wastes and by-products come from several industries, as the agri-food industry. These resources present an enormous potential for valorization in the textile finish due to their intrinsic properties (antimicrobial, prebiotic, antioxidant activity, among others). This chapter will review the latest innovation and textile product development through different by-products and wastes, their main properties and characteristics and the advantages that they offer to the textile industry.

Keywords: innovative textiles, functional textiles, waste valorization, sustainability, circular economy, protein fraction, agro-industrial wastes, vegan leather, leather waste, alternative leather

1. Introduction

In the textile industry, the utilization of low environmental impact technologies that are based on sustainable raw materials presents a novel possible way for the development of functional textiles on a large scale. By-products and wastes from different sources and industries such as proteins, vegetable, agroforestry, furniture, food, footwear and automotive industries are often used as biomass or sent to landfills. However, due to the interest in by-products as a substituent for the commercially aggressive chemicals used in the textile industry, research on the valorization of these materials has remarkably increased. In this sense, several studies were carried out to enhance the performance attributes of textile goods through finishing, coating and dipping technologies with by-products and residues, thus creating an opportunity for the establishment of partnerships and circular economy business models.

2. Natural by-product and waste compositions and properties

2.1 Whey protein

The dairy industry is characterized by a broad group of food products, such as milk, milk powder, butter, yoghurts, cream and cheese, but it is also a big source of solid and liquid by-products, but among those, whey is the one produced at the highest volumes in cheese industry. The world production of by-products in dairy industry is around 4–11 million tonnes per year, but Europe is the worldwide leader in cheese production and consequently the largest whey producer [1, 2]. This has a big environmental impact if they are disposed as wastes, so strategies to reuse these by-products are important, and there is a community pressure in this sense. Traditionally, some years ago, whey is used to be disposed of, but with environmental concerns and legislation to be implemented, the reuse appears with a prominent role [2]. Whey is considered one of the major pollutant by-products because of its high biological and chemical oxygen demands [2]. Whey is composed of 85–90% water, 10–15% lactose (carbohydrates), soluble vitamins, minerals (e.g. calcium, phosphorus, sodium and so on) and proteins (e.g. β-lactoglobulin, α-lactalbumin, bovine serum albumin (BSA), immunoglobulins and others) [1, 2]. Lactose is the main component, being responsible for most of the biological and chemical oxygen demands [2]. However, lactose and other nutrients essential for microbial growth confer whey a potential to produce several bioproducts. Whey proteins award health benefits such as high nutritional value, easy digestion and assimilation, which are interesting for the food industry too. It can be used for biotransformation feeds, bioproteins, prebiotics, and bioactive peptides after fermentation or enzymatic hydrolysis. On the other hand, the reduced-lactose whey, demineralized whey, and whey protein concentrates or isolates are used for food, cosmetic and pharmaceutical industries, especially for thier emulsifying, thickening, gelling, foaming and water-binding properties. More specific, these proteins of natural origin and with emulsifying capacity are used in the formulation of creams and shampoos as substitutes for synthetic surfactants. The whey protein hydrolysate also has this type of application for hair products. Another property of whey protein consists of gel formation, being used to produce protective films and coatings. These proteins, as they have low permeability to water vapour, are used in paper coating, providing good appearance and printability. β-Lactoglobulin and α-lactalbumin can be used as moisturizing and antiwrinkle agents. Lactoferrin is a good iron chelator, preventing the formation of free radicals. In the 1990s, whey protein, in the form of iron proteinate, was also used as an antianemic preparation [3]. Moreover, this by-product can be reused not only for its technological properties but also for its biological properties in terms of the body's benefits.

Due to its biological and chemical properties, whey has several applications, depending on the biotechnological method applied to reuse this by-product. It can be used to obtain chemical products to produce functional formulations and for food, fuel health, pharmaceuticals, biomaterials and others.

Whey has been used in the food industry to produce functional food and drinks as an innovative product with health benefits. For example, whey can be used in dairy beverages: unfermented or fermented, probiotic, refreshing soft, alcoholic, diet and high protein sport [2]. Whey is also one of the main sources of the bioactive peptides that can be used as nutritional supplements because of its structure, rapid absorption and biological properties (antihypertensive activities and antioxidant properties). On the other hand, whey protein has been explored in the food industry in edible film or coating development for food preservation, for its biochemical properties, such as its edible nature and intrinsic biodegradability, suitable mechanical barrier, flexibility and the capacity to incorporate functional compounds [4].

The improvement of knowledge has also allowed the development of biomaterials from whey, for example, to produce biodegradable capsules for drug delivery. The whey protein isolates (WPI) have been used for bone regeneration to produce bioactive glasses with potential applications in bone tissue engineering.

These biomaterials have been explored for tissue engineering applications due to their chemical and biological properties, such as the ability to retain water, easy transport/entrapment of nutrients or cells, controlled biodegradability, mechanical properties and biocompatibility [5]. However, the field of action of whey is much wider and applied to other industries as an eco-friendly alternative to conventional chemicals.

2.2 Agro-industrial wastes

Agro-industrial wastes include several different wastes from the food and agriculture industries. The amount of wastes from the food and forestry-based industries produced in the European Union (EU) is estimated to be in the order of 900 million tonnes per year. However, a large part of these wastes are considered low-value input materials instead of wastes, like sawdust that can be used to make products such as fibreboard or leaves and stalks of plants that can have other agricultural uses such as animal bedding [6]. If these wastes are released to the environment without a proper disposal procedure, they may worsen the environmental pollution and cause harmful effects on human and animal health. **Table 1** shows the estimated sustainable availability of agro-industrial wastes.

Recently, these wastes have been the focus of much attention due to their huge potential for exploration, not only for their wide availability and diversity but also for their intrinsic properties and functionalities, which make them an increasingly

Wastes	Current availability (Mtonnes/year)	2030 availability (Mtonnes/year)
Paper industry	17.5	12.3
Wood industry	8	5.6
Food and garden industries	37.6	26.3
Crop wastes	122	139
Forestry wastes	40	40

Table 1.
Agro-industrial wastes and wastes produced in the EU [6].

attractive feedstock for chemical, material and biofuel production [7]. Conscious consumption allied with ethical and sustainable values is increasing the consumers' concern in the moment of purchase: "What is the nature of the raw material?"; "What is the life cycle of the product?". This tendency has made the producers look for alternative raw material sources [8].

It was found that the typology of vegetable wastes most produced varies from year to year, with the most abundant being materials unsuitable for consumption or processing, biodegradable wastes and vegetable textile wastes [9]. The most promising vegetable and agroforestry wastes for textile application are, for example, sawdust, coffee grounds, pine bark, eucalyptus bark and others. Sawdust and composites of sawdust (in powder and in pieces) are very abundant wastes as a result of the wood processing industry such as furniture industry. Coffee grounds are highly abundant because the cultural habit of people is drinking a lot of coffee. Pine bark is a highly abundant waste that is very easy to adapt for textile coating applications, which can result in a brown powder that gives rise to coatings with a dark colour and a very attractive shade. Olive stones are also abundant, resulting from the production of olive oil or from the ginning of the olives. Almond or nutshell wastes can create coatings with very attractive colours and visual effects. Rice husk, due to its low nutritional value, is not a viable resource as food for animals, and the burning or landfill deposition of this type of waste has important environmental impacts, as it has a slow biological degradation (high silica content). Eucalyptus bark is also abundant, resulting from the paper and wood processing industries [7].

2.3 Leather wastes

The transformation of animal skins into leather allows for the recycling of what would be an organic waste from the food industry into added-value products. In this context, the animal skin is considered a by-product, as it is not reintroduced in the same productive cycle and its reuse contributes to a more sustainable and a circular economy.

There are several applications for leather, and the manufacture of leather upholstery for furniture, airplanes and automobiles has been one of the main markets in the last two decades. Although leather waste recycling has been the subject of hundreds of studies, landfilling remains the most frequent option, wasting all resources contained in leather. Also, due to environmental restrictions, the study and development of sustainable alternatives for the recovery of this waste for the manufacture of new, more sustainable materials are urgent [10].

The valorization of leather wastes such as leather shavings aims to the reduction of the presence and usage of Cr (VI), oil, hydrocarbon, and solvent absorber; adsorbent of chlorides, fats, tannins, surfactants, and dyes, used in the tanning process. Leather powder has already been applied as an oil and crude absorber, while carding powder has been used as an adsorbent for textile dyes (more anionic than cationic) [10].

This type of waste can be physically processed by crushing and grinding methods. For certain uses, its mixture with resins and catalysts for subsequent pressing between metal moulds with various configurations and sizes can produce multilayer or composite structures. Final products are obtained with a very good appearance, without the need for any additional finishing, with good sound insulation and even good thermal insulation [11]. Applications in furniture, floors and footwear components are some of the examples. Through these processes leather wastes have been used in leather-like materials and construction materials, as additives for thermoplastic composites and as filler materials for reinforcing rubbers [10].

Leather waste can also be processed chemically (alkaline or acid hydrolysis) or enzymatically, in order to obtain collagen (by-product) for application in added-value products. Collagen consists of a fibrous, insoluble and inert protein, which after alkaline/acid/enzymatic hydrolysis is divided into gelatine and hydrolysed (soluble) collagen, by breaking the chromium-collagen bond established during the tanning phase and breaking non-covalent bonds in the protein's structure that lead to its swelling and solubilization [12, 13].

The chemical processing of leather wastes also results in Cr (VI), which can be reintroduced upstream into the leather tanning process. Another type of chemical processing reported for the recovery of Cr (VI) involves the incineration of tanned chips and blue chips and later transformation of the ashes by converting chromium (III) oxide into sodium chromate [Cr (VI)] [14, 15].

Given that the present method of recovering collagen from leather wastes is free of complex installations and equipment, its implementation in the productive cycle of companies is economically attractive [14].

3. Functional properties of natural by-products and wastes

3.1 Whey protein

Whey exhibits many unique functional properties such as antibacterial and antioxidant activity and odour and water vapour absorber, among others. Therefore, whey has become an attractive product for its versatile applications in different fields, including textile industry. Many of these applications are also reported in the development of new functional products in the food and pharmaceutical fields, due to the properties (such as antimicrobials, antioxidants, and anticancer drugs) and structures of whey protein and its fractions. **Table 2** shows some examples of applying these fractions to obtain the functionalities described.

Another application for whey or milk fractions is related to the production of microcapsules. In fact, globular proteins had been used as a vehicle for the micro-/nanoencapsulation of bioactive compounds. Milk proteins, namely, whey protein, have been used for the microencapsulation of aromas. Using serum protein isolate and gum arabic, it is possible to encapsulate β-carotene. The same gum arabic had already been shown to be effective in promoting self-aggregation, and consequent capsule formation, of β-lactoglobulin [28–30]. Another aspect is the microencapsulation of β-lactoglobulin with another polysaccharide, chitosan, and this has a

Functionality	Description	Ref.
Antioxidant	Several studies show that whey has antioxidant properties. It is maximized with an enzymatic treatment of whey, milk or cheese and with the hydrolysate's valorization (microbial proteases, β-lactoglobulin and α-lactalbumin). This evaluation was done with ABTS or ORAC-FL method	[16–19]
Deodorant property	Milk and whey proteins are effective in the absorption of odours, given their composition in proteins and lipids. Lactose is described by its ability to retain odours, absorbing them on its surface as the crystals form	[20, 21]
Antimicrobial	Two of the whey fractions, lactoferrin and lactoperoxidase, present an antimicrobial activity. Lactoferrin has several antimicrobial peptides that are released after hydrolysis by proteases. Lactoperoxidase has a high antimicrobial capacity through catalytic and chemical processes	[22–27]

Table 2.
Whey, protein fraction and dairy by-product functionalities.

stabilizing effect on serum proteins, protecting them from denaturation at temperatures up to 90°C. Due to its structure, β-lactoglobulin can also form complexes with vitamins and nutraceuticals, such as folic acid. β-Lactoglobulin/folic acid complexes exhibit particle sizes below 10 nm and exhibit stability over a wide range of pH values [31–33].

3.2 Agro-industrial wastes

The passage of traditional industrial processes to more sustainable patterns and a circular economy model are mandatory given the limited resources and adverse environmental effects that are noticeable today. In this sense, the establishment of bio-based economies and industrial processes, such as the textile industry, will contribute directly to substitute emission-intensive and non-renewable resources with renewable resources, as well as create innovative and functional added-value solutions [9]. Some wastes or natural additives can provide a wide range of functional properties to textiles, opening an opportunity for the development of new and innovative textile solutions. Some potential functionalities of some vegetable and agroforestry wastes and by-products are presented in **Table 3**.

3.3 Leather wastes

The manufacture of leather upholstery for furniture, airplanes and automobiles has been one of the main markets in the last two decades. Currently, in Europe, 14% of all new cars have leather coverings, and an additional 4% are made in combinations of leather, textiles, composite materials and imitation leather. The world's leading car manufacturers have focused on looking for renewable materials, recycling materials in manufacturing processes and using less toxic materials to improve car recyclability [43]. In the European footwear industry, the production of about $1–2 \times 10^5$ tonnes of leather waste per year is estimated, with the annual cost associated with its management between 4 and 10×10^6 € [44]. In the manufacture of footwear, more than 70% of the leather used is leather tanned with chromium [10].

Despite the many methodologies and systems studied and implemented in the last decades, which allowed the minimization of waste production during the manufacture of leather and its processing by user industries, such as the automotive and footwear industries, these production processes inevitably generate waste leather which can be disposed or valorized as it is or by chemical conversion into other added-value products (collagen) [10].

Native collagen and its derivatives are widely applied in the food, agrarian (fertilizer), cosmetic and biomedical industries, as well as in the textile industry, due

Waste/by-product	Source	Functionalities	Ref.
Coffee grounds	Coffee production process	Anti-odour, antimicrobial, aromatic; UV radiation protection	[34, 35]
Rice husks	Rice processing	Thermal insulation potential	[36]
Eucalyptus bark	Wood processing industry	Antimicrobial, aromatic	[37, 38]
Pine bark	To feed	Antioxidant, antimicrobial, aromatic	[39, 40]
Pine sawdust, composite sawdust, powder and pieces	Wood processing industry	Absorbent, mechanical and structural properties	[41, 42]

Table 3.
Vegetable and agroforestry wastes and by-products and functionalities.

Functionality	Description	Ref.
Gelling and dilating	Aggregation of molecules at 30°C to form hydrolysed collagen gels and gelatine; swelling in the presence of water	[45–47]
Foaming	The presence of hydrophobic and hydrophilic amino acids provides excellent foaming properties, even in the absence of gelling	
Antimicrobial	Hydrophobic amino acids penetrate the peptide chains that make up bacterial membranes, acting as a natural fungicide and bactericide	
Antioxidant/ anti-ageing	Inhibition of lipid peroxidation, elimination of free radicals and acting as transition metal ion chelating agents, protecting cells from damage caused by oxidation and helping to improve skin firmness	

Table 4.
Different functionalities of collagen and its derivatives and respective area of application.

to their biodegradability, biocompatibility, etc. [15]. In addition, collagen and its derivatives have also another set of properties that enhance their potential, not only for the direct functionalization of textile substrates but also for the development of the coating formulations (**Table 4**) [15, 45, 46].

4. Functional applications of natural by-products and wastes in the textile industry

The consumer demand for more environmentally responsible products with better sustainability credentials is increasingly growing, in addition to progressively more restrictive legislation regarding the environmental impact of industrial activity. Additionally, other increasingly important factors are the search for textile products with differentiated technical and functional properties and with better sustainability credentials, without compromising the appearance, touch, and comfort of the article.

These facts have led companies in the textile and clothing sector to gradually invest in an investigation strategy that leads to the adoption of sustainable policies and reduction of environmental impacts, based on the valorization of wastes and by-products of industries that are geographically close. In this scenario, the reuse of these natural by-products and wastes as a bio-resource in the demanding textile sector presents itself as an alternative.

4.1 Textile fibres with whey protein

The use of milk proteins for fibre production and application in textile industry remotes back to the beginning of the twentieth century. The conventional fibre production method consists in dissolving 20–25% milk proteins, including whey protein and its fractions, in a 2% NaOH solution to obtain a solution of adequate viscosity for fibre production by wet spinning extrusion (10–30% solid material) [48, 49]. In this process, the protein solution is pumped through a spinneret into an acid bath with a pH below the isoelectric point of the protein (4.5–4.6) to cause its coagulation [48, 50, 51]. The coagulate is afterwards stretched and drawn to increase polymer chain orientation and tensile strength of the fibre. Coagulation baths, containing aluminium salts of formaldehyde, may further increase the fibre stretching and enhance its physical properties [48, 51].

There are already several studies and patents on the production of fibres from whey proteins aiming to obtain fibres with improved mechanical properties and to use of more ecological productive processes. Kamada et al. produced fibres from β-lactoglobulin nanofibrils in the presence of alcohols, low pH and elevated temperature (hydrolysis of the protein in low molecular weight peptides for the formation of nanofibrils) [52]. Sullivan et al. produced nanofibres, by electrospinning, from WPI solutions (75%) and polyethylene oxide (PEO) (4%) and solutions of β-lactoglobulin (75%) and PEO (10%) in water [53]. Drosou et al. [54] studied the possibility to make whey protein fibres by electrospinning. However, electrospinning of nanofibres from proteins has proven to be quite challenging due to their globular nature, in most cases, the low viscosity of their aqueous solutions and potential lack of intermolecular entanglements [54]. To overcome these challenges, blends of proteins and other bio-based materials have been used. Drosou also tested some WPI/pullulan blends and was able to obtain continuous and uniform fibres [54]. The presence of the pullulan increased the viscosity of the solution, having a big impact in the process parameters. Zhong et al. adopted a similar strategy to obtain also whey protein nanofibres through electrospinning [55]. In this case the authors blended the whey protein with PEO and were not able to produce pure protein fibres. The ability of the whey protein solutions to produce fibres changed over time after dissolution [55]. Oktar et al. produced fibres from WPC blended with poly-ε-caprolactone (80 kDa) [56]. The obtained fibres showed improved mechanical properties to higher WPC concentrations (3–8% w/v). Kutzli et al. produced whey protein fibres by electrospinning, blending the proteins with enzymatically treated starch (maltodextrin) [57]. Using two different maltodextrins, with different molecular weights, the authors found that the spinnability of the solution is heavily dependent on the average size of the maltodextrin. Aman Mohammadi et al. obtained whey protein fibres by electrospinning, mixing WPI and guar gum [58].

As already mentioned, fibres resulting from these processes usually fail to have the mechanical properties for weaving and textile production. For this reason, whey protein fibres are often mixed with other fibres with appropriate mechanical properties (mostly cotton, silk and wool, with tensile strengths) [59].

The valorization of by-products of the dairy industry by wet spinning generates corrosive effluents rich in metal salts. This type of effluent requires appropriate conditioning and downstream steps of neutralization and precipitation of metals, which may entail large costs for its treatment and disposal (in order to avoid acidification of soils and water resources, increase of the dissolved salt content and the appearance of health problems in animals and humans resulting from untreated discards in water bodies used to supply populations) [60].

4.2 Textile finishing with whey protein

Whey proteins have also been studied for their applicability as coatings and additives in the textile industry. Pisitsak et al. (2015) studied the dyeability increase of cotton for a tannin-rich dye extracted from *Xylocarpus granatum* bark. Cotton fabrics were pretreated with WPI by a padding technique. The improvement in the dye absorption after protein pretreatment is ascribed to the insoluble complex formation between the tannin and the proteins present in the fabric, stabilized through hydrogen bonding and hydrophobic interactions, which makes it easy to be coloured. Besides that, both protein treatment and dyeing improved the ultraviolet (UV) shielding efficiency of the cotton fabrics [61].

Proteins are not the only milk component able to facilitate the dyeing process. Dyes are generally applied in an aqueous solution, and some of them require chemical auxiliaries to improve their water solubility and to improve the dyeing process. Bianchini et al. [62] reported a study to naturalize two synthetic azadyes through their linkage with lactose to induce their water solubility. In this study, a chromophore was transformed into a hydrosoluble species through glycol conjugation with a sugar, and a preliminary tinctorial test was carried out with polyester, cotton, acetate, wool and acrylic fabrics. Results showed several benefits since the modification of the dyes with lactose, as this improved their water solubility, allowing the elimination of surfactants and mordants, making the dyeing process easier and avoiding high temperatures and high pressures. Besides that, the new hydrosoluble dyes showed a better affinity towards different fabrics (synthetic, natural, artificial), improving efficacy and reducing waste [62].

These developments brought benefits not only in terms of textile valorization but also in terms of the use and recovery of wastes and by-products. The utilization of carbohydrates largely and cheaply available, such as D-glucose, D-galactose and lactose, normally discarded in huge quantities in the environment, with no negligible impact, brings new possibilities for efficient and more selective waste treatment by using, for instance, live micro-organisms to attack the sugar moiety and consequently the covalently bonded chromophore, or the use of enzymes able to destroy dyes [62].

In the past years, novel and innovative solutions for flame retardant systems, for replacing the traditional additives, have been explored. In particular, the availability of a formaldehyde-free flame retardant system based on natural macromolecules such as proteins could be extremely interesting for a possible industrial application [63]. Considering the environmental concern, more ecological and effective solutions have been studied, in the field of flame retardancy, since the solutions mostly used are based on halogenates or phosphorus, being persistent and bioaccumulating in the soil and even carcinogenic and/or toxic for animals and humans. In this sense, biomacromolecules have aroused interest as a green solution in this field, particularly whey proteins and caseins. In addition to being biological additives, they can have added value, as they can be considered by-products or even wastes from the agro-food industry and their recoveries and subsequent use as flame retardants may comply with the current needs of valorization of agro-food crops, avoiding their landfill confinement [57, 58].

Therefore, different novel strategies have been designed in order to enable the use of green flame retardant systems. Due to the ability of whey proteins to act as water vapour absorbers and as oxygen barriers, textiles treated with this by-product have been exploited in order to increase their thermal stability and flame retardancy [63]. For this, folded and unfolded whey protein isolates were deposited on cotton fabrics. Through thermogravimetric analysis it was observed that whey protein coatings significantly affected the thermal degradation of cotton in an inert and oxidative atmosphere. Specifically, the application of whey protein coating contributed to the delay of the thermal degradation of the textile, also resulting in a smaller total mass loss. Besides that, the treated fabrics have shown a decrease of burning rate and an increase of total burning time, determined by the flammability tests in horizontal configuration [63].

The antibacterial properties of some of the whey components have also been studied. Through the cross-linking between microbial transglutaminase (mTGase) and lactoferrin, the antibacterial properties of wool were improved to *E. coli* (Gram-negative) and *S. aureus* (Gram-positive) bacteria. It was observed that the amount of lactoferrin deposited on the wool fabric was improved with the cross-linking reaction with mTGase, when compared to the control sample. The wool fabrics immobilized with lactoferrin exhibited approximately 70 and 60% inhibition for *E. coli* and *S. aureus*, respectively, showing a good antibacterial property [64].

The same was observed in a recent study developed by Srisod et al. [65]. It was described the utilization of WPI as reducing and stabilizing agent in a green synthesis of silver nanoparticles (AgNps) from silver nitrate. In addition, a natural tannin-rich extract was applied to cross-link the WPI/AgNps to cotton fabric through the formation of an insoluble binder. The cotton fabric treated showed an excellent antibacterial performance against *S. aureus*, even after 50 washing cycles, showing no toxicity to L929 cell changes to the intrinsic properties of the substrate (drapeability and tearing strength) [65].

Regarding the globular structure of whey proteins, due to their properties and structures, they have been used as a vehicle for active substances such as antimicrobials, antioxidants and drugs, among others, for the development of new functional products [66–69]. This approach is widely used in several industrial sectors, providing the possibility of a controlled release of bioactive compounds. It can easily be applied to the textile industry, with the possibility to add functionality to textiles.

The antioxidant effects of vitamin E encapsulated in BSA nanoparticles in cotton have already been studied [70]. The nanoparticles, produced by ultrasonic emulsification, have a size between 200 and 300 nm and have the capacity to encapsulate 99% of the vitamin. After impregnation onto cotton fabrics, they present an antioxidant activity and wash resistance up to ten cycles [71].

Microspheres of BSA have also been tested as encapsulation agents of an antibiotic, tetracycline, in order to obtain an antibacterial coating for cotton and polyester fabrics [72]. These capsules demonstrated not only good encapsulation capacity but also gave the textiles antimicrobial properties [72].

Nonetheless, these types of applications at an industrial level have some limitations since the cost-effectiveness ratio of these biomacromolecules may not compensate until now. In addition, the durability to the laundering was not yet achieved in an effectively sustainable and long-lasting way, since these biomacromolecules have a waterborne character and these coatings come off from the textile when subjected to washing. When adding binding agents to biomacromolecules, a balance must be sought between their green characteristics and the use of chemicals that do not eradicate the sustainability of the process. In this sense, exploitation of biologically derived chemical treatments, or at least chemicals with a low environmental impact, which could make the proposed biomacromolecules more durable than they are today, while maintaining their effective functionalities, is being carried out [73, 74].

4.3 Textile coating as a sustainable alternative to genuine leather

Genuine leather is made of animal skin, namely, bovine leather, tanned and finished with products of synthetic origin (chromium). It is used as a noble material for the manufacture of various products with applications in various industries, such as fashion, fashion accessories, footwear, decoration, automobiles, etc., and is the one that has the greatest expression in the market due to its excellent properties such as porosity, breathability, softness, comfort and fall, among others [75, 76]. Ecological leather refers to a leather tanning process that does not use metals such as chromium but in alternative recurs to substances of natural origin (vegetable, animal or mineral), such as vegetable tannins (polyphenols of plant origin) [77]. Though ecological leather has a lesser environmental impact than genuine leather, it still does not have the same properties of thermal resistance, colour fixation and versatility as the leather resulting from the treatment of tanning with chromium [75]. In addition, there are several ethical and environmental concerns involved in the use of genuine and ecological leather, such as the killing of animals and the high environmental impact resulting from their processing, which have triggered the growing interest on the part of the consumer in more sustainable alternative solutions to leather of animal origin ethically and environmentally. This generated a search for alternative solutions with the same performance of genuine leather, which catapulted textile industries towards sustainable innovation as a means of answering the markets' demands.

Vegetable leather is a sustainable product of plant origin resulting from the use of vegetable wastes or by-products. There are already some alternatives of vegetable leather on the market to replace animal leather, although they do not fully reproduce the characteristics of animal leather. Of the solutions on the market, the main examples are presented.

Latex-based leather is the name given to a fabric made up of two renewable raw materials, the latex extracted from the rubber tree (*Hevea brasiliensis*) from the Amazonian forests and cotton. The cotton is impregnated with latex, natural rubber (primary product of the smoking of latex extracted from the rubber tree). These can be used in the production of bags, wallets, clothing, footwear and other objects usually produced in leather. The commercialization of these products has become a reason for hope for the improvement of the life of rubber tappers, their permanence in the forest and the sustainable development of the Amazon, generating work and income in indigenous and traditional communities [78–80].

The company Ananas Anam has developed an innovative, natural and sustainable non-woven leather called Piñatex™, produced from pineapple leaf fibres, considered as a vegan alternative to traditional leather. From the pineapple leaf fibres, screens are obtained, which can be dyed, printed and treated to obtain different textures [81]. The material is strong, versatile (different colours, patterns, textures, thicknesses), breathable, smooth, light, flexible, sewable, resistant to water and abrasion and resistant to ignition by cigarettes [82–84].

Products based on thin sheets of cork, laminated with a textile substrate that gives it resistance, are increasingly being introduced to the market as a sustainable vegan alternative to traditional/synthetic leather. They have characteristics equivalent to leather, such as resistance, lightness, breathability, malleability, thermal insulation and impermeability, adding the properties of low density and thermal conductivity. There are several products based on cork leather (cork sheet) on the market, created and launched by designers/brands and national reference companies, such as Bleed—We bleed for nature, Pelcor, and Artelusa, and international, such as Chanel, inter Louboutin, Stella McCartney, Yves Saint Laurent, Prada, Dior, Manolo Blahnik, Dolce & Gabbana and Gucci. These products are based on fashion accessories (wallets, belts, etc.), clothing, umbrellas, footwear, sports goods, furniture, car upholstery lining, etc. [76, 85–90].

Wood-based leather is similar to cork but made from wood from fast-growing trees, such as oak bark, treated with non-toxic chemicals to make it durable, flexible and malleable. Wood leather can be as thick as genuine leather. Dolce & Gabbana is a market reference that has already used this material in a recent collection of bags and shoes [41]. The German shoe brand nat-2™ also recently launched a line of shoes in which up to 90% of the upper surface of the shoe is covered with wood, which is applied over an organic cotton, in order to become a flexible, soft material that allows to smell the wood and observe its natural texture [91]. Another solution is Wooden Textiles, created by Elisa Strozyk. These materials, which also bear some resemblance to leather, are obtained after cutting thin sheets of wood into pieces and adhering them to a textile substrate. The result is a material that smells like wood, but with some flexibility and softness. There are applications in decoration and furniture [92].

Vegea® is a biomaterial produced by the Italian company Vegea, founded by Gianpiero Tessitore and Francesco Merlino [86, 87]. This material, with a similar aspect to leather, valorizes residues from bagasse (skins and tales from grapes), and does not use water in its production [74]. This leather, also known as WineLeather, is already available in several colors, and it can be used for studying or obtaining different thicknesses, strengths, finishes, and textures. It is already applied in the production of clothing, bags and shoes, furniture, packaging, and automobile and transport accessories [93]. It is used to coat a textile substrate with a polymeric mixture, consisting of a cake residue flour and a derived polymer of oil extracted from grapes [94].

The German company nat-2™ developed a material similar to leather, obtained from coffee bean wastes [95, 96]. With this material a line of unisex sneakers was created, whose upper part contains recycled coffee, coffee beans and coffee plant, which constitutes up to 50% of the footwear surface, according the model. The coffee is applied in a layer, giving a soft touch and a coffee aroma. Two Mexican inventors, Adrian Lopez and Marte Cazarez, recently created a laminate based on nopal cactus (or figs), which resembles animal leather, that is breathable, environmentally sustainable and totally plant-based (cotton and Nopal blend), lasts at least 10 years and has the chemical and physical properties required by the fashion industries, furniture, leather goods and automobiles [97, 98]. The material is obtained by coating a cotton substrate with a mixture of dry (in the sun) and crushed cactus powder and protein extracted from the cactus, which serves as a natural binder [99].

Another leather-like material example is bonded leather or reconstituted leather. This consists of the preparation of a paste with ground leather wastes and binding agents, which is extruded, using a process similar to the production of paper [100]. This paste can be applied on a textile support, coated with a PU film and embossed to gain a leather-like texture [101]. The colour and pattern are checked by a surface treatment. The amount of leather fibres in bonded leather can vary, which is reflected in the quality of the material. This product is usually used in furniture, bookbinding and fashion accessories. Depending on the quality of the product, it can be a durable material, with flame retardancy, and does not develop a patina. The number of patents on reconstituted or recycled leather is extensive, without, however, mentioning the use of textile support for the application of the paste with leather wastes [102–124].

RecycLeather™ is a green technology company that produces recycled materials with the look and feel of leather, highly durable, resistant and light. The materials are obtained from leather waste, in particular, cut pieces from gloves. It consists of 60% leather waste, 30% latex (a natural binder) and 10% synthetic products, such as water and pigments [125].

EcoDomo also has some collections with recycled leather [126]. This is obtained by pulverized leather fibres, obtaining materials with a leather content of up to 70%. It is available for different applications, such as furniture, panels, flooring, etc. Embrace™ also has different materials, similar to leather, obtained from leather waste (43–58%), blended with cotton and polyester, and a PU topcoat [127].

Hydrolysed collagen has recently been applied in the leather manufacturing process, and in the production of flexible composite sheets, with polyvinylpyrrolidone (PVP) and cellulose derivatives, for application products in the area of footwear, clothing, etc. [128–130]. The application of collagen hydrolysates in leather production consisted of its mixture with oxazolidines before application, but the obtained results were not as good as those attained by tanning [45, 46]. The application of this by-product, without chromium separation, in the manufacture of flexible composite sheets with both PVP and cellulose allowed the obtaining of composites with improved mechanical properties (composites with PVP and cellulose) and greater thermal stability (cellulose composites) [47, 131].

Gelatex is a non-woven fabric (with nanofibres) made from gelatine derived from waste from the meat and leather industries, developed by Gelatex Technologies, a start-up from Estonia [132]. It is a material with a touch similar to leather and is breathable, durable and customizable (texture, thickness, water resistance, etc.). This material won the The Green Alley Award 2019 [133].

5. Case of study

The mobilizing project TexBoost—less Commodities more Specialties is a structuring project of the Textile Cluster: Technology and Fashion, which aims to include a set of R&D initiatives with a strong collective character and high inductor and demonstrator effect, with the central involvement of companies of the textile and clothing sector, but also of other complementary sectors of the economy [134]. TexBoost consortium, led by RIOPELE and under the technical coordination of CITEVE, involves a total of 43 entities, of which 23 are industrial companies of the entire textile industry and 15 are non-corporate entities of the research and innovation system.

The project is organized into six PPS—products, processes and services—of which it is worth highlighting the PPS5, sustainability and circular economy. This PPS5 aims the development of materials and solutions using wastes and by-products of other industries (footwear, automobile, cork, forest and milk industry) in new and innovative textile solutions.

For the first nuclear activity, vegan leather, the R&D work was focused in the development of a new generation of coated textile solutions that could be used as an alternative to natural and/or synthetic leather, using wastes and by-products of vegetable origin with new multifunctional properties combined with design and special fashion effects form the basis of this activity. The aim of this work were also to respond to one of the major trends in consumption, related to ethically and environmentally sustainable attitudes, developing products with a high potential for application in technical and functional areas, such as technofashion, eco-design, clothing, decoration, home textiles, footwear, fashion accessories, sport and protection, among others.

During the project, several agro-industrial wastes were studied, and from them, eco-friendly and Registration, Evaluation, Authorisation and Restriction of Chemicals (REACH)-compliant coating formulations were developed, and 100% cotton textile substrates were coated by knife coating (**Figure 1**).

The mechanical performance of the developed solutions was assessed through a series of normalized tests, namely, Veslic friction resistance (ISO 11640: 2012), Martindale abrasion resistance (ISO 17704:2004), Crockmeter friction resistance (ISO 20433:2012), colour fastness (ISO 105-B02) and coating peeling (ISO 11644:2009) (N/cm). The obtained results are summarized in **Table 5**. In a general way, it is possible to conclude that the developed solutions pass the performance norms and specifications.

Regarding the second nuclear activity—alternative leather solutions—the R&D activities focused on the development of a new generation of coated textile solutions by using wastes and by-products resulting from industrial operations, such as the tanning industry, natural leather cutting (for indoor automotive) and EVA (for shoe components), here highlighting the leather wastes, with new multifunctional properties combined with fashion design and special effects. The aim was also to meet one of the major trends of current consumption, which is related to ethically and environmentally sustainable behaviour, developing products with high potential for application in technical and functional areas and in rapid expansion: technofashion, eco-design, clothing, decoration, home textiles, footwear, fashion accessories, sport and protection, among others.

During the project, leather waste was studied, eco-friendly and REACH-compliant coating formulations were developed, and 100% cotton textile substrates were coated by knife coating (**Figure 2**).

Figure 1.
Vegan leather solutions based on sawdust (left) and coffee grounds (right).

Normative test	Coffee ground-based vegan leather	Sawdust-based vegan leather
Veslic[a,b]	5	5
Martindale[c]	3200 rev.: A 6400 rev.: B 51,200 rev.: B	3200 rev.: A 6400 rev.: B 12,800 rev.: B 25,600 rev.: C 51,200 rev.: C
Crockmeter[d]	5	5
Colour fastness[e]	3–4	3
Coating peeling (dry)/(N/cm)	30	31.2

[a]*Veslic friction resistance: flower side, degree of staining; dry skin/wet felt—50/100 cycles.*
[b]*Veslic friction resistance: flower side, degree of colour change; dry skin/wet felt—50/100 cycles.*
[c]*Martindale abrasion resistance: dry, abrasion degree.*
[d]*Crockmeter friction resistance: flower side, degree of staining; wet and dry—ten cycles.*
[e]*Light fastness: xenon lamp, flower side; colour fastness.*

Table 5.
Mechanical performance evaluation of the coffee ground-based vegan leather and of the sawdust vegan leather samples.

The mechanical performance of the developed solutions was assessed through a series of normalized tests, namely, Veslic friction resistance (ISO 11640: 2012), Martindale abrasion resistance (ISO 17704:2004), Crockmeter fiction resistance (ISO 20433:2012), and colour fastness (ISO 105-B02). The obtained results are summarized in **Table 6**. In a general way, it is possible to conclude that the developed solutions pass the performance norms and specifications.

Finally, the other approach of the PPS was research and development of a new generation of coated textile solutions, using wastes and by-products of the dairy industry, with new multifunctional properties combined with design and special fashion effects. Specifically, the two main goals were functionalization of textiles with milk proteins to improve UV protection and use of milk proteins to encapsulate bioactive compounds (such as antioxidants) and subsequent functionalization of textiles.

So, in the present project, 2,2-azino-bis-(3-ethylbenzothiazoline)-6-sulfonic acid (ABTS) method was used for the evaluation of antioxidant activity of a whey protein fraction impregnated in textile substrate. This spectrophotometric method assesses the stabilization capacity of the ABTS radical formed from certain compounds. In other words, it indicates the percentage of inhibition of the ABTS radical after contact with the compounds.

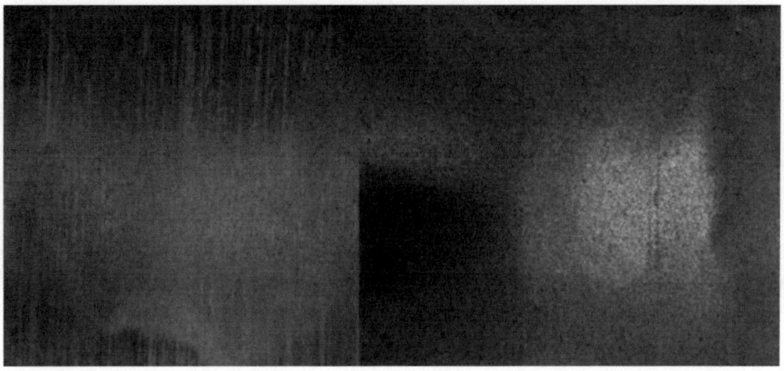

Figure 2.
Alternative leather solutions based on leather waste (left) and hydrolysed collagen (right).

Normative test	Leather waste-based alternative leather	Hydrolysed collagen-based vegan leather
Veslic[a,b]	3–5	3–5
Martindale[c]	1600 rev.: A 3200 rev.: B 12,800 rev:B 25,600 rev:C 51,200 rev:C	1600–3200 rev.: A 6400–51,200 rev.: B
Crockmeter[d]	2–5	3–5
Colour fastness[e]	3–4	4–5

[a]*Veslic friction resistance: flower side, degree of staining; dry skin/wet felt—50/100 cycles.*
[b]*Veslic friction resistance: flower side, degree of colour change; dry skin/wet felt—50/100 cycles.*
[c]*Martindale abrasion resistance: dry, abrasion degree.*
[d]*Crockmeter friction resistance: flower side, degree of staining; wet and dry—ten cycles.*
[e]*Light fastness: xenon lamp, flower side; colour fastness.*

Table 6.
Mechanical performance evaluation of the alternative leather samples.

For this, microcapsules of a milk fraction were prepared with and without an antioxidant compound. These microcapsules were used to functionalize a textile substrate and analysed by ABTS method. To the textile substrate, the relative antioxidant ability to scavenge the radical ABTS+ was compared to the textile control, without functionalization. It was possible to verify that all the protein fraction gave the substrates significantly higher ABTS inhibition percentages than the controls, with a slight increase when the antioxidant is present.

Since textiles had a high antioxidant potential, the capacity of this potential was verified in terms of protecting the colours of textiles when exposed to UV radiation. In this way, the textiles were stained with a dye and exposed for 12 hours to UV radiation. It was found that after 12 hours of exposure to UV radiation, the control showed a high degradation of the stain colour. On the other hand, the functionalization of textiles delayed the process of colour photodegradation, since after 12 hours of exposure, none of the stains had yet reached the same colour reduction.

6. Conclusions

The potential for reusing natural by-products and wastes from different sources was reviewed in this chapter, describing their most attractive properties and characteristics. The most recent innovations and developments in this area were listed and presented, showing a novel possible way for the development of technical and functional textiles. The main potential applications for the valorization of whey protein by the production of textile fibres have been described, as well as by its application as a textile finish. The different applications already tested and the main products already available on the market for sustainable alternatives to produce genuine leather were also listed. Although these types of applications at an industrial level have some limitations, as cost-effectiveness ratio, permanence of the intrinsic properties of the substrates and durability to the laundering, for example, the reuse of these natural by-products and wastes as a bio-resource in the demanding textile sector presents itself as an attractive alternative.

Acknowledgements

The mobilizing project TexBoost—less Commodities more Specialties (no 24523), in PPS 5, sustainability and circular economy, a project co-financed by COMPETE 2020—Operational Program for Competitiveness and Internationalization—and in Portugal 2020 through the European Regional Development Fund (ERDF).

Author details

Lorena Coelho[1*], Ana Isabel Magalhães[1], Sara Fernandes[1], Patrícia Batista[2], Manuela Pintado[2], Pedro Faria[1], Catarina Costa[1], Bruna Moura[1], Augusta Marinho[3], Rosa Maria[3], Albertina Reis[4], Marta Carvalho[4], Mário Marques[4], Ângela Teles[4], José De Almeida Morgado[3], Maria Helena Vilaça[3], Jéssica Alexandra Pereira[3], Pedro José Magalhães[5], Ana Sofia Silva[5], Ricardo Jorge Silva[5], Mário Jorge Silva[5], Vera Lúcia Sá[6], Sandra Gabriela Ventura[7], João Silva Abreu[7], Joaquim Manuel Gaião[8], Raquel Rosa Mourão[8], Fernando Manuel Merino[9], Mónica Sofia Gonçalves[9] and Regina Malgueiro[1]

1 Centre of Nanotechnology and Smart Materials (CeNTI), Vila Nova de Famalicão, Portugal

2 Associated Laboratory of Faculty of Biotechnology of the Catholic University of Portugal (ESB-UCP), Centre of Biotechnology and Fine Chemistry (CBQF), Porto, Portugal

3 Technological Centre for the Textile and Clothing Industries of Portugal (CITEVE), Vila Nova de Famalicão, Portugal

4 Riopele—Têxteis SA, Av. Pousada de Saramagos, Portugal

5 Tintex—Textiles SA, Vila Nova de Cerveira, Portugal

6 Sedacor-Sociedade Exportadora De Artigos De Cortiça Lda, São Paio de Oleiros, Portugal

7 Têxteis Penedo, SA, Mascotelos, Portugal

8 Technological Centre for the Leather Industry (CTIC), Alcanena, Portugal

9 Ert Têxtil Portugal, SA, São João da Madeira, Portugal

*Address all correspondence to: lcoelho@centi.pt

References

[1] Ahmad T, Aadil RM, Ahmed H, Rahman U, Soares BCV, Souza SLQ, et al. Treatment and utilization of dairy industrial waste: A review. Trends in Food Science and Technology. 2019;**88**:361-372. DOI: 10.1016/j.tifs.2019.04.003

[2] Papademas P, Kotsaki P. Technological utilization of whey towards sustainable exploitation. Advances in Dairy Research. 2020;7(4):1-10. DOI: 10.35248/2329-888x.7.4.231

[3] Audic J-L, Chaufer B, Daufi G. Non-food applications of milk components and dairy co-products: A review. EDP Sciences. 2003;**83**:417-438. DOI: 10.1051/lait

[4] Fernandes LM, Guimarães JT, Pimentel TC, Esmerino EA, Freitas MQ, Carvalho CWP, et al. Edible whey protein films and coatings added with prebiotic ingredients. In: Agri-Food Industry Strategies for Healthy Diets and Sustainability. Cambridge: Academic Press; 2020. pp. 177-193. DOI: 10.1016/b978-0-12-817226-1.00007-2

[5] Gupta D, Kocot M, Tryba AM, Serafim A, Stancu IC, Jaegermann Z, et al. Novel naturally derived whey protein isolate and aragonite biocomposite hydrogels have potential for bone regeneration. Materials & Design. 2020;**188**:108408. DOI: 10.1016/j.matdes.2019.108408

[6] Searle S, Malins C. Availability of Cellulosic Residues and Wastes in the EU [Internet]. Washington, USA: International Council on Clean Transportation; 2013. Available from: https://theicct.org/sites/default/files/publications/ICCT_EUcellulosic-waste-residues_20131022.pdf

[7] Sadh PK, Duhan S, Duhan JS. Agro-industrial wastes and their utilization using solid state fermentation: A review. Bioresources and Bioprocessing. 2018;**5**(1). DOI: 10.1186/s40643-017-0187-z

[8] Agência Portuguesa do Ambiente. Plano De Acção De Qualidade Do Ar [Internet]. Available from: https://apambiente.pt/index.php?ref=17&subref=152

[9] Thorenz A, Wietschel L, Stindt D, Tuma A. Assessment of agroforestry residue potentials for the bioeconomy in the European Union. Journal of Cleaner Production. 2018;**176**:348-359. DOI: 10.1016/j.jclepro.2017.12.143

[10] Ferreira MJ. Valorização De Resíduos De Couro Curtido Com Crómio Provenientes Da Indústria Do Calçado. CTCP, Projeto Financiado Pelo QREN No Âmbito Do SIAC: São João Da Madeira; 2012

[11] What Is Recycled Leather? A Guide to Recycled Leather [Internet]. Available from: https://www.brighthub.com/environment/green-living/articles/91891.aspx

[12] Vaskova H. Hydrolysis of leather shavings: Mathematical model and experiment. In: Katalinic B, editor. Proceedings of the 29th DAAAM International Symposium. Vienna, Austria: DAAAM International; 2018, pp. 0338-0340. DOI: 10.2507/29th.daaam.proceedings.048. ISBN: 978-3-902734-20-4, ISSN: 1726-9679

[13] Mohammad AW, Suhimi NM, Aziz AGKA, Jahim JM. Process for production of hydrolysed collagen from agriculture resources: Potential for further development. Journal of Applied Sciences. 2014;**14**(12):1319-1323. DOI: 10.3923/jas.2014.1319.1323

[14] Processo Para Reciclagem De Resíduos De Couro. Escavador

[Internet]. Available from: https://www.escavador.com/patentes/43747/processo-para-reciclagem-de-residuos-de-couro

[15] Moreira S. Estudo Da Obtenção De Gelatina A Partir De Raspa Wet-Blue Da Indústria De Curtumes. Instituto Politécnico do Porto, Instituto Superior de Engenharia do Porto; 2008

[16] Corrêa APF, Daroit DJ, Fontoura R, Meira SMM, Segalin J, Brandelli A. Hydrolysates of sheep cheese whey as a source of bioactive peptides with antioxidant and angiotensin-converting enzyme inhibitory activities. Peptides. 2014;**61**:48-55. DOI: 10.1016/j.peptides.2014.09.001

[17] Dryáková A, Pihlanto A, Marnila P, Čurda L, Korhonen HJT. Antioxidant properties of whey protein hydrolysates as measured by three methods. European Food Research and Technology. 2010;**230**(6):865-874. DOI: 10.1007/s00217-010-1231-9

[18] Hernández-Ledesma B, Dávalos A, Bartolomé B, Amigo L. Preparation of antioxidant enzymatic hydrolysates from α-lactalbumin and β-lactoglobulln. Identification of active peptides by HPLC-MS/MS. Journal of Agricultural and Food Chemistry. 2005;**53**(3):588-593. DOI: 10.1021/jf048626m

[19] Liu HC, Chen WL, Mao SJT. Antioxidant nature of bovine milk β-lactoglobulin. Journal of Dairy Science. 2007;**90**(2):547-555. DOI: 10.3168/jds.S0022-0302(07)71538-2

[20] Lasztity R. Milk and milk products. In: Lasztity R, editor. Food Quality and Standards. Vol. 2. Hungary: Encyclopedia of Life Support Systems; 2009. pp. 156-167

[21] Pomeranz Y. Functional properties of food components. In: Food Science and Technology. 2nd ed. Cambridge: Academic Press; 1991

[22] Sinha M, Kaushik S, Kaur P, Sharma S, Singh TP. Antimicrobial lactoferrin peptides: The hidden players in the protective function of a multifunctional protein. International Journal of Peptides. 2013;**390230**:1-12. DOI: 10.1155/2013/390230

[23] Kussendrager KD, van Hooijdonk ACM. Lactoperoxidase: Physico-chemical properties, occurrence, mechanism of action and applications. British Journal of Nutrition. 2000;**84**(S1):19-25. DOI: 10.1017/s0007114500002208

[24] Bafort F, Parisi O, Perraudin J-P, Jijakli MH. Mode of action of lactoperoxidase as related to its antimicrobial activity: A review. Enzyme Research. 2014;**2014**:517164. DOI: 10.1155/2014/517164

[25] Weight KC, Tramee J. Factors influencing the activity of cheese starters: The role of milk peroxidase. Journal of Dairy Research. 1958;**25**(1):104-118. DOI: 10.1017/S0022029900009080

[26] Portmann A, Auclair JE, Gaye Y. Relation entre La lacténine L 2 Et La lactoperoxydase. Le Lait. 1959;**39**(383-384):147-158. DOI: 10.1051/lait:1959383-3846

[27] Yener FYG, Korel F, Yemenicioğlu A. Antimicrobial activity of lactoperoxidase system incorporated into cross-linked alginate films. Journal of Food Science. 2009;**74**(2):M73-M79. DOI: 10.1111/j.1750-3841.2009.01057.x

[28] Koupantsis T, Pavlidou E, Paraskevopoulou A. Flavour encapsulation in milk proteins E CMC coacervate-type complexes. Food Hydrocolloids. 2014;**37**:134-142

[29] Jain A, Thakur D, Ghoshal G, Katare OP, Shivhare US. Microencapsulation by complex coacervation using whey protein

isolates and gum acacia: An approach to preserve the functionality and controlled release of β-carotene. Food and Bioprocess Technology. 2015;**8**(8):1635-1644. DOI: 10.1007/s11947-015-1521-0

[30] Sanchez C, Mekhloufi G, Schmitt C, Renard D, Robert P, Lehr CM, et al. Self-assembly of B-lactoglobulin and acacia gum In aqueous solvent: Structure and phase-ordering kinetics. Langmuir. 2002;**18**:10323-10333

[31] Chen L, Subirade M. Chitosan/β-lactoglobulin core–shell nanoparticles as nutraceutical carriers. Biomaterials. 2005;**26**(30):6041-6053. DOI: 10.1016/j.biomaterials.2005.03.011

[32] Zhao Z, Xiao Q. Effect of chitosan on the heat stability of whey protein solution as a function of PH. Journal of the Science of Food and Agriculture. 2017;**97**(5):1576-1581. DOI: 10.1002/jsfa.7904

[33] Pérez OE, David-Birman T, Kesselman E, Levi-Tal S, Lesmes U. Milk protein–vitamin interactions: Formation of beta-lactoglobulin/folic acid nano-complexes and their impact on In vitro gastro-duodenal proteolysis. Food Hydrocolloids. 2014;**38**:40-47. DOI: 10.1016/j.foodhyd.2013.11.010

[34] Singh A. A novel approach of coffee groud fiber towards environment friendly textile. IMPACT. 2019;**7**(12):29-32

[35] Almeida AA, Farah A, Silva D, Nunan E, Gloria MBA. Antibacterial activity of coffee extracts and selected coffee chemical compounds against enterobacteria. Journal of Agricultural and Food Chemistry. 2006;**54**:8738-8743. DOI: 10.1021/jf0617317

[36] Muthuraj R, Lacoste C, Lacroix P, Bergeret A. Sustainable thermal insulation biocomposites from rice husk, wheat husk, wood fibers and textile waste fibers: Elaboration and performances evaluation. Industrial Crops and Products. 2019;**135**:238-245. DOI: 10.1016/j.indcrop.2019.04.053

[37] Jain P, Nimbrana S, Kalia G. Antimicrobial activity and phytochemical analysis of eucalyptus tereticornis bark and leaf methanolic extracts. International Journal of Pharmaceutical Sciences Review and Research. 2010;**4**(2):126-128

[38] Ali S, Nisar N, Hussain T. Dyeing properties of natural dyes extracted from eucalyptus. Journal of The Textile Institute. 2007;**98**(6):559-562. DOI: 10.1080/00405000701556079

[39] Karakaya PS, Oktay A, Seventekin N, Yesil-Celiktas O. Design of a new generation wound dressing with pine bark extract. Journal of Industrial Textiles. 2019;**0**:1-12. DOI: 10.1177/1528083719855324

[40] Chan C, Goyal A, Hung O, Islam S, Jajpura L, Kan C, et al. In: Nayak R, editor. Sustainable Technologies for Fashion and Textiles. Woodhead Publishing Series in Textiles. UK: Woodhead Publishing; 2020. DOI: 10.1016/B978-0-08-102867-4.01002-1

[41] Vignesh V, Balaji AN, Karthikeyan MKV. Effect of wood sawdust filler on the mechanical properties of Indian mallow Fiber yarn mat reinforced with polyester composites. International Journal of Polymer Analysis and Characterizatio. 2017;**22**(7):610-621. DOI: 10.1080/1023666X.2017.1356481

[42] Badu M, Boateng I, Boadi N. Evaluation of adsorption of textile dyes by wood sawdust. Research Journal of Physical and Applied Science. 2014;**3**:6-14

[43] Mao Z, Jin Y. Reverse Logistics in Automotive Industry. 2014

[44] Salden MP. Chromium: Environmental, Medical and Materials Studies. Hauppauge: Nova Science Publishers; 2011. pp. 1-384

[45] INSUMOS. Aditivos & Ingredientes—Parte I. Revista No. 63 ARTIGO 2 [Internet]. 2009. Available from: https://aditivosingredientes.com.br/artigos/ingredientes-funcionais/colageno-entenda-o-que-e-i

[46] INSUMOS. Aditivos & Ingredientes—Parte II. Revista No 63 ARTIGO 2 [Internet]. 2009. Available from: https://aditivosingredientes.com.br/artigos/ingredientes-funcionais/colageno-entenda-o-que-e-ii

[47] Gomez-Guillen MC, Gimenez B, Lopez-Caballero ME, Montero MP. Functional and bioactive properties of collagen and gelatin from alternative sources: A review. Food Hydrocolloids. 2011;**25**(8):1813-1827. DOI: 10.1016/j.foodhyd.2011.02.007

[48] Fire E. What Is Milk Fiber. 2011

[49] Gupta V, Kothari V. Manufactured Fibre Technology. Dordrecht: Springer Netherlands; 1997. pp. 133-135. DOI: 10.1007/978-94-011-5854-1

[50] Pomeranz Y, Pomeranz Y. Some Traditional Foods: Functional Properties of Food Components. Elsevier; 1991. pp. 381-473. DOI: 10.1016/B978-0-12-561281-4.50014-1. Available from: http://linkinghub.elsevier.com/retrieve/pii/B9780125612814500141

[51] Macrae R, Robinson R. Encyclopedia of Food Science, Food Technology, and Nutrition. Michigan: Academic Press; 1993

[52] Kamada A, Mittal N, Söderberg LD, Ingverud T, Ohm W, Roth SV, et al. Flow-assisted assembly of nanostructured protein microfibers. Proceedings of the National Academy of Sciences of the USA. 2017;**114**(6):1232-1237. DOI: 10.1073/pnas.1617260114

[53] Sullivan ST, Tang C, Kennedy A, Talwar S, Khan SA. Electrospinning and heat treatment of whey protein nanofibers. Food Hydrocolloids. 2014;**35**:36-50. DOI: 10.1016/j.foodhyd.2013.07.023

[54] Drosou C, Krokida M, Biliaderis CG. Composite pullulan-whey protein nanofibers made by electrospinning: Impact of process parameters on fiber morphology and physical properties. Food Hydrocolloids. 2018;**77**:726-735. DOI: 10.1016/j.foodhyd.2017.11.014

[55] Zhong J, Mohan SD, Bell A, Terry A, Mitchell GR, Davis FJ. Electrospinning of food-grade nanofibres from whey protein. International Journal of Biological Macromolecules. 2018;**113**:764-773. DOI: 10.1016/j.ijbiomac.2018.02.113

[56] Oktar FN, Su S, Ozbek B, Yücel S, Kazan D, Gunduz O. Production and characterization of whey protein concentrate (WPC) based nano-fibers. Materials Science Forum. 2018;**923**:47-50. DOI: 10.4028/www.scientific.net/MSF.923.47

[57] Kutzli I, Gibis M, Baier SK, Weiss J. Fabrication and characterization of food-grade fibers from mixtures of maltodextrin and whey protein isolate using needleless electrospinning. Journal of Applied Polymer Science. 2018;**135**(22):1-9. DOI: 10.1002/app.46328

[58] Aman Mohammadi M, Ramazani S, Rostami M, Raeisi M, Tabibiazar M, Ghorbani M. Fabrication of food-grade nanofibers of whey protein isolate–guar gum using the electrospinning method. Food Hydrocolloids. 2019;**90**:99-104. DOI: 10.1016/j.foodhyd.2018.12.010

[59] Domaske A. Process for Producing Milk Protein Fibers and Milk Protein Fiber Products Obtained Therefrom [Internet]. US20130256942A1, 2013.

Available from: https://patents.google.com/patent/US20130256942A1/en

[60] Guimarães D, Lopes KCS, Rodrigues MLM, Bertolino SM, Leão VA. Tratamento de efluentes ácidos ricos em metais com auxílio de sulfeto biogênico e cal. HOLOS. 2014;**3**:332. DOI: 10.15628/holos.2014.1812

[61] Pisitsak P, Hutakamol J, Thongcharoen R, Phokaew P, Kanjanawan K, Saksaeng N. Improving the dyeability of cotton with tannin-rich natural dye through pretreatment with whey protein isolate. Industrial Crops and Products. 2016;**79**:47-56. DOI: 10.1016/j.indcrop.2015.10.043

[62] Bianchini R, Catelani G, Frino E, Isaad J, Rolla M. Lactose to naturalize textile dyes. BioResources. 2007;**2**(4):630-637

[63] Bosco F, Carletto RA, Alongi J, Marmo L, Di Blasio A, Malucelli G. Thermal stability and flame resistance of cotton fabrics treated with whey proteins. Carbohydrate Polymers. 2013;**94**(1):372-377. DOI: 10.1016/j.carbpol.2012.12.075

[64] Han X, Yu Y, Wang Q, Fan X, Cui L, Wang P. Anti-bacterial properties of Lactoferrin immobilized wool fabric. Indian Journal of Fibre & Textile Research. 2014;**39**(4):401-405

[65] Srisod S, Motina K, Inprasit T, Pisitsak P. A green and facile approach to durable antimicrobial coating of cotton with silver nanoparticles, whey protein, and natural tannin. Progress in Organic Coatings. 2018;**120**(October 2017):123-131. DOI: 10.1016/j.porgcoat.2018.03.007

[66] Chen Q, McGillivray D, Wen J, Fang Zhong SQ. Co-encapsulation of fish oil with phytosterol esters and limonene by milk proteins. Journal of Food Engineering. 2013;**117**:505-512

[67] de Kruif CG(K), Weinbreck F. Complex coacervates of whey proteins and anionic polysaccharides, and their use for encapsulation. Food New Zealand. 2005;**5**:23-30

[68] Augustin MA, Augustin MA, Sanguansri L, Oliver CM. Functional properties of milk constituents: Application for microencapsulation of oils in spray-dried emulsions—a mini review. Dairy Science & Technology. 2010;**90**:137-146

[69] Moreau DL, Rosenberg M, Miller MM, Ziegler G. Microstructure and fat extractability in microcapsules based on whey proteins or mixtures of whey proteins and lactose. Food Structure. 1993;**12**(4):457-468

[70] Ghaheh FS, Khoddami A, Alihosseini F, Jing S, Ribeiro A, Cavaco-Paulo A, et al. Antioxidant cosmetotextiles: Cotton coating with nanoparticles contanining vit E. Process Biochemistry. 2017;**59**:46-51

[71] Ghaheh FS, Khoddami A, Alihosseini F, Gomes A, Ribeiro A, Cavaco-Paulo A, et al. Protein-based nanoformulations for α-tocopherol encapsulation. Engineering in Life Sciences. 2016;**17**(5):523-527

[72] Shimanovich U, Cavaco-Paulo A, Nitzan Y, Gedanken A. Sonochemical coating of cotton and polyester fabrics with "antibacterial" BSA and casein spheres. Chemistry--A European Journal. 2012;**18**(1):365-369. DOI: 10.1002/chem.201100781

[73] Malucelli G, Bosco F, Alongi J, Carosio F, Di Blasio A, Mollea C, et al. Biomacromolecules as novel green flame retardant systems for textiles: An overview. RSC Advances. 2014;**4**(86):46024-46039. DOI: 10.1039/c4ra06771a

[74] Malucelli G. Biomacromolecules and bio-sourced products for the design of flame retarded fabrics: Current state of the art and future perspectives.

Molecules. 2019;**24**(20):3774, 1-27. DOI: 10.3390/molecules24203774

[75] de Almeida AI. Novos Processos E Produtos Para O Couro De Base Vegetal [Internet]. Instituto Politécnico do Porto. Instituto Superior de Engenharia do Porto; 2011. Available from: http://recipp.ipp.pt/handle/10400.22/4550

[76] Silva A, Peixoto J, Souto AP. Development and Application of a New Concept of Cork Substrate In Footwear and Clothing. 2013. pp. 1-6

[77] Magma Têxtil [Internet]. Available from: https://www.lojamagma.com.br/

[78] Manzoni WA. Process and Composition for Obtaining a Rubbery Tissue [Internet]. 2001. Available from: https://patents.google.com/patent/WO2002036874A1

[79] de Brandão MLF. Design Sustentável: O Uso Da Matéria Prima Renovável. Um Estudo De Caso Da Produção Do Couro Vegetal No Norte Do Brasil. 2007

[80] Programa de Aceleração (PPA). Encauchados De Vegetais Da Amazônia [Internet]. 2020. Available from: http://aceleracao.ppa.org.br/portfolio-de-negocios/encauchados/

[81] Vegan Leather Options and Alternatives to Animal Leather [Internet]. Available from: https://ecowarriorprincess.net/2016/07/vegan-leather-options-alternatives-to-animal-leather/

[82] Piñatex [Internet]. Available from: https://www.ananas-anam.com/

[83] Conheça O Couro Vegano Feito De Abacaxi [Internet]. Available from: https://www.noticiasaominuto.com.br/lifestyle/274656/conheca-o-couro-vegano-feito-de-abacaxi

[84] Inovador Tecido Natural Feito Da Folha Do Abacaxi [Internet]. Available from: https://www.divaholic.com.br/design/couro-vegano-feito-de-abacaxi/

[85] 7 Sustainable Vegan Textiles you Should Know About [Internet]. Available from: https://theminimalistvegan.com/sustainable-vegan-textiles/

[86] Sustainable Clothing—Fair & Vegan|Bleed [Internet]. Available from: https://www.bleed-clothing.com/english/

[87] 15+ Eco Friendly Vegan Leather Alternatives—Eluxe Magazine [Internet]. Available from: https://eluxemagazine.com/fashion/5-truly-eco-friendly-vegan-leathers/

[88] Silva Â, Souto AP. Estudo Da Substituição Do Couro Pela Pele De Cortiça Para Aplicação Em Calçado/Vestuário [Internet], 2013. Available from: http://repositorium.sdum.uminho.pt/handle/1822/37039

[89] Vegan Cork Jackets Become Germany's Most Successful Fashion Campaign! Veg [Internet]. Available from: https://www.clearlyveg.org/blog/2016/02/02/vegan-cork-jackets-become-germanys-most-successful-fashion-campaign/

[90] Sewell C. From Apples to Kombucha Tea: See the Ingenious Way Designers Are Making Vegan Leather [Internet]. People for the Ethical Treatment of Animals. 2015. Available from: https://www.peta.org/living/personal-care-fashion/vegan-leather-chic-sustainable-and-fruity/

[91] The World's First Vegan Wooden Sneakers by Nat-2™ [Internet]. Available from: https://nat-2.eu/collections/vegan-wooden-sneakers/

[92] Elisa Strozyk—Wooden Textiles [Internet]. Available from: www.elisastrozyk.de/seite/woodentextiles.html

[93] VEGEA [Internet]. Available from: https://www.vegeacompany.com/company/

[94] Periodic Reporting for Period 1—VegeaTextile (Innovative Biomaterials Production from Wine Industry Waste) [Internet]. Available from: https://cordis.europa.eu/project/id/805055/reporting

[95] New Vegan Sneaker Range by Nat2 Is Made out of Coffee Leather [Internet]. Available from: https://www.livekindly.co/nat-2-vegan-sneaker-range-coffee-leather/

[96] Nat-2. Nat-2™ Coffee Line [Internet]. 2018. Available from: https://nat-2.eu/collections/nat-2-coffee-line/

[97] Inventores Mexicanos Criam Material Alternativo Ao Couro Feito De Cacto Nopal [Internet]. Available from: https://www.stylourbano.com.br/inventores-mexicanos-criam-material-alternativo-ao-couro-feito-de-cacto-nopal/

[98] Jovens Mexicanos Criam Inovadora Pele Orgânica Feita De Cacto: Adeus Ao Couro Animal! [Internet]. Available from: https://razoesparaacreditar.com/pele-organica-cacto/

[99] Mexican Entrepreneurs Invent Leather Alternative based on Cactus [Internet]. Available from: https://www.youtube.com/watch?v=gc43eRWe07U

[100] Types of Leather Furniture [Internet]. Available from: https://www.myfurniture.com.sg/101/leather-furniture-types.php

[101] For Consumers' Sake, Let's Not Call it "Bonded Leather" [Internet]. Available from: https://www.furnituretoday.com/business-news/for-consumers-sake-lets-not-call-it-bonded-leather/

[102] Adolph AG. Leather Substitute and like Composition [Internet]. 1922. Available from: https://patents.google.com/patent/US1527163

[103] Charles B. Process of Manufacturing Reconstituted Compressed Leather and Product Obtained Thereby [Internet]. US2040511A, 1934. Available from: https://patents.google.com/patent/US2040511A

[104] Young Harland H, Edward H Nahja RHE. Reconstituted Leather Product and Method of Making [Internet]. US3116200A, 1957. Available from: https://patents.google.com/patent/US3116200A

[105] Harland H Young, Edward J Majka RHE. Method for Producing Leather Fiber Slurry [Internet]. US3179342A, 1962. Available from: https://patents.google.com/patent/US3179342A

[106] Albert E Raymond, William J, Fraser FSJ. Water-Laid Leather Substitute Containing Leather Fibers,Staple Fibers and Polyurethane Aqueous Slurry and Method for Making Same [Internet]. US3436303A, 1965. Available from: https://patents.google.com/patent/US3436303A

[107] Michael Barash BSA. Method of Making Fibrous Sheet Material [Internet]. US3542910A, 1966. Available from: https://patents.google.com/patent/US3542910A

[108] Parker ET. Reconstituted Leather and Method for Producing it [Internet]. US3505169A, 1966. Available from: https://patents.google.com/patent/US3505169A

[109] Parrini P, Peroni G, Corrieri G RG. Fibrous Materials Useful As Leather Substitutes and Consisting Essentially of Leather Fibers, Fibrils or Fibrides of Synthetic Polymers and Cellulose Fibers [Internet]. US4162996A, 1976. Available from: https://patents.google.com/patent/US4162996A

[110] Dimiter S. Reconstituted Leather and Method of Manufacture [Internet]. US4287252A, 1980. Available from:

https://patents.google.com/patent/US4287252A

[111] Tsui Y-M. Process for Manufacturing Regenerated Leather [Internet]. US4325236A, 1980. Available from: https://patents.google.com/patent/US4325236A

[112] Picagli RG, Tucker ES, Infantino JR, Light HJ, Marinaccio JVF PJ. Reconstituted Leather Product from Fibrillated Leather Fibers [Internet]. 1983. Available from: https://patents.google.com/patent/EP0089029A2

[113] Incorporated AMF, Picagli RG, Tucker ES, Infantino JR, Light HJ, Marinaccio JVF PJ. Reconstituted Leather Product from Fibrillated Leather Fibers [Internet]. 1983. Available from: https://patents.google.com/patent/EP0089029A2

[114] Henke EW. Reconstituted Leather and Method of Manufacturing Same [Internet]. US4497871A, 1983. Available from: https://patents.google.com/patent/US4497871A

[115] Pelzer H. Moulded Articles with Leather-like Surface Properties for Use In the Automobile Industry [Internet]. US5624619A, 1992. Available from: https://patents.google.com/patent/US5624619A

[116] Addie BA. Reconstituted Leather Product and Process [Internet]. US5958554A, 1996. Available from: https://patents.google.com/patent/US5958554A

[117] Benjamin A. Addie JK. Reconstituted Leather Product and Process [Internet]. US6264879B1, 1999. Available from: https://patents.google.com/patent/US6264879B1

[118] Suese H. Manufacturing Reconstituted Leather from Slaughterhouse Waste, Chrome Leather Waste and Mineral Fibers, Employs Tanning, Degreasing, Liming and Pressure-Swing Processing [Internet]. DE10060246A1, 2000. Available from: https://patents.google.com/patent/DE10060246A1

[119] Han S. Production of Regenerated Leather by Dry Method [Internet]. US20040149369A1, 2001. Available from: https://patents.google.com/patent/US20040149369A1

[120] Coulson N, Homan Kinsley JN. Composite Leather Material [Internet]. US20070184742A1, 2006. Available from: https://patents.google.com/patent/US20070184742A1

[121] Rosa OD. Regenerated Bonded Leather and Method for Making it [Internet]. WO2012001490A1, 2010. Available from: https://patents.google.com/patent/WO2012001490A1

[122] 朱晓华. Low-Cost Collagen Fiber Reconstituted Leather and Manufacturing Method Thereof [Internet]. CN103233324A, 2013. Available from: https://patents.google.com/patent/CN103233324A

[123] Bevan CG. Formation of Sheet Material Using Hydroentanglement [Internet]. US20140113520A1, 2001. Available from: https://patents.google.com/patent/US20140113520A1

[124] Benjamin M. Rappoport JAS. Composite Bonded Leather Cases [Internet]. US20150360442A1, 2014. Available from: https://patents.google.com/patent/US20150360442A1

[125] Recyc Leather™. Sustainable Recycled Leather [Internet]. Available from: http://www.recycleather.com/

[126] EcoDomo—Leather Surfacing & Fabrication Company [Internet]. Available from: http://ecodomo.com/

[127] Embrace Recycled Leather [Internet]. Available from: https://

fabricsupply.com/products/embrace-recycled-leather/

[128] Afşar A, Gülümser G, Aslan A, Ocak B. A study on usability of collagen hydrolysate along with oxazolidine in leather processing. Tekstil Ve Konfeksiyon. 2010;**20**(1):37-40

[129] Ashokkumar M, Thanikaivelan P, Murali R, Chandrasekaran B. Preparation and characterization of composite sheets from collagenous and chromium-collagen complex wastes using polyvinylpyrrolidone: Two problems, one solution. Waste and Biomass Valorization. 2010;**1**(3):347-355. DOI: 10.1007/s12649-010-9030-x

[130] Ashokkumar M, Thanikaivelan P, Krishnaraj K, Chandrasekaran B. Transforming chromium containing collagen wastes into flexible composite sheets using cellulose derivatives: Structural, thermal and mechanical investigations. Polymer Composites. 2011;**32**(6):1009-1017. DOI: 10.1002/pc.21121

[131] Huang L, Nagapudi K, Apkarian RP, Chaikof EL. Engineered collagen–PEO nanofibers and fabrics. Journal of Biomaterials Science. Polymer Edition. 2001;**12**(9):979-993. DOI: 10.1163/156856201753252516

[132] Gelatex [Internet]. Available from: https://www.gelatex.com/

[133] Estonian Startup Gelatex Wins Green Alley Award 2019 [Internet]. Available from: https://www.baltictimes.com/estonian_startup_gelatex_wins_green_alley_award_2019/

[134] TexBoost—Home [Internet]. Available from: https://www.texboost.pt/

Chapter 7

Value Addition to Leather Industry Wastes and By-Products: Hydrolyzed Collagen and Collagen Peptides

Ali Yorgancioglu, Bahri Başaran and Aykut Sancakli

Abstract

Environmental consciousness and constraints in developed societies over the past 20 years have brought about a dramatic impact on tannery operations worldwide. Leather industry has been categorized as one of the most polluting industries, and it spoils the continuity of environmental rhythm because of the generation of liquid, solid and gaseous wastes and also by-products. Solid organic wastes involving untanned (trimmings, fleshings and splits) and tanned (trimmings, splits and shavings) wastes and by-products depending on their proteinic character have an advantage of recovery and reuse potentials instead of disposal to landfills in terms of environmental sustainability. These solid wastes and by-products are not properly treated and disposed of; hence, they can cause environmental damages to soil and groundwater as well as release emissions and poisonous greenhouse gases into the atmosphere. Valorization of these tannery solid wastes and by-products with different methods and processes is highly important for the perspective of eco-benignity and with respect to converting into new value-added products. This chapter focuses on the evaluation of the tannery solid wastes and by-products by partial and total denaturation and hydrolyzation. This paper also examines in general the specifications, production techniques and applications of collagen peptides in several industries.

Keywords: leather, solid wastes and by-products, technical gelatin, collagen peptides, collagen hydrolysates

1. Introduction

Leather manufacturing, which is an allied industry and subsector for textile, is the first making practice in primitive period of humankind. Different types of animal skin products were used throughout the first ages as parchment and vellum or by making the raw material resistant to putrefaction, heat, chemicals and environmental effects with smoke, potash alum and natural tannin extracts from different plant parts. Traditionally, these products obtained by modification of by-products of meat industry have all been classified as leather, which is a serviceable product. In this respect, the leather industry could have been distinguished as an environmental industry, since it processes waste products from meat production [1].

IntechOpen

These natural products generally consist of long thick collagen fibers, fiber bundles and thin elastin fibers of interweaving in three-dimensional ways. Other features such as hairs and hair roots and also fat cells are present in three-dimensional woven structure that predominates and gives skin-based materials providing many of their unique physical and mechanical qualities [2].

The leather-making operation assists in converting the raw hide or skin, a highly putrescible material, into leather, a stable material, which can be used in manufacturing a wide range of products. These include shoes, clothing, leather goods, furniture, upholstery for car seats and interiors, boats and aircraft, and many other goods in daily use. The whole process involves a sequence of complex chemical reactions and mechanical processes [3].

The processing of leather involves four main stages: beamhouse, tanning, post-tanning and finishing. The first phase of the hide processing is called beamhouse operations and involves multiple mechanical, chemical and biological unit operations. Its objective is to remove dirt, hair, epidermis, noncollagenous proteins and grease from raw skin, and open up the collagen fibers to favor the subsequent tanning process [4]. The process is performed in a drum by mixing the raw hides with an alkaline solution containing lime and reducing agents, usually sulfide salts, the hair being chemically removed from the surface of the hide [5]. The beamhouse operations are the most water consuming and the effluents generated present very high organic load [6].

The tanning process is one of the oldest procedures in the world, and currently, these industrial activities are based on chemical processes involving several organic and inorganic compounds [7]. This step gives the leather stabilization against the wet and dry heat, bacterial growth, mechanical stress and enzymatic attack, among others, and forms the basis of leather production. This stabilization is attributed to the formation of new chemical cross-links in the matrix proteins [8]. The tanning stages are classified as mineral, vegetable and synthetic. When the skin stabilization is achieved by a suitable inorganic salt, the process is known as mineral tanning, and the most commonly used mineral tanning salt is the basic chromium sulfate ($Cr(OH)SO_4$). If the leather is tanned with chromium salt, it is called as wet-blue leather. Chromium (III) salts are the most extensively used compounds due to the quality and high stabilization ability they impart to leather [9].

The third part in leather production is post-tanning process. The tanned leather is considered a commodity, that is, it may be used to produce several articles. Each post-tanning operation is directed to the article that will be produced, such as garment, shoe upper and upholstery [10]. The aim of the post-tanning processes is to enhance the aesthetic properties of leather by coloring it and changing some physical and mechanical properties of the material by retanning, dyeing and fatliquoring stages [10].

The finishing step complements the previous stage, tanning, and provides the leather with the required physical and mechanical properties, such as color, tensile strength, impermeability, softness, flexibility and elasticity with different kinds of binder, pigment, wax and oils [11]. This operation consists of coating and changing the surface of leather. It is related to the fashion appearance, but also to conferring properties such as abrasion resistance, gloss, handle, flex, adhesion and rub fastness as well as other properties as required for the end use including extensibility, light and perspiration fastness, water vapor permeability and water resistance [10].

Leather industry has been categorized as one of the highly polluting industries because large quantities of water and different chemicals have been used during tanning process and different solid, gaseous and liquid wastes are generated that have an adverse effect on the environment [12]. These wastes have different characteristics because different chemicals are applied to the raw hides in different ratios. Solid wastes generated in tanneries mainly include salts, raw trimmings, hair wastes, fleshings, splitting wastes, chrome shavings, buffing dusts, crust trimmings and finished trimmings. These solid wastes and by-products are not properly treated and disposed of, and they can cause environmental damages to soil and groundwater as well as release emissions of odor and poisonous greenhouse gases into the atmosphere by direct landfill or incineration, which is an unsustainable way [13].

Salt, which is used to preserve hides or skin thrown into open dumping areas or accumulated in piles outside the tanneries, is likely to create groundwater pollution when rain washes it away. Hair wastes and lime sludge discharged into the effluent can produce choking of treatment pipelines. Trimmings, fleshings and splitting wastes putrefy easily producing noxious odors [14]. Moreover, disposal of chromium-containing solid wastes into soil and water has potential effects on public health due to the possibility of oxidation of chromium (III) into hazardous chromium (VI) [15]. These tannery solid wastes have different characteristics that mainly constitute protein (collagen) as the main component [16].

Provisions for pollution control, waste minimization and disposal, the correct use of chemicals and accident prevention are essential for minimizing potential impact on air, water and soil from the processing of hides and skin.

Collagen derivatives are value-added products extracted from solid organic wastes and by-products, and they are utilized for several industrial applications such as preparation of technical-grade gelatin, protein hydrolysates, collagen peptides and subunits [17]. The processing of hides and skin also generates by-products, which find outlets in several industrial sectors such as pet and animal food production. They can be used in cosmetics, printing inks and photography, while the latter one is an ideal candidate for fertilizer or feeding additives due to their high nitrogen content [18].

The present chapter describes the leather solid wastes, general features of collagen peptides, and their preparation methods and applications in different industries.

2. Leather solid wastes

The tanning industry worldwide produces a significant amount of solid wastes and effluents, environmental concerns about discharge and escalating landfill costs are becoming increasingly serious problems for the industry, and their management alternatives regarding overall consideration have been based on multispot [19]. Huge amounts of solid wastes are generated at different stages of leather processing and there is no actual adopted utilization method available for solid wastes; hence, handling is more difficult for tanners. Leather solid wastes generated in fleshing, trimming, splitting and shaving processes and also sludges discharged from the wastewater treatment plant both contribute to increase the volume of the wastes [20].

Generally, out of 1000 kg of rawhide, nearly 800 kg of solid wastes are generated in leather-manufacturing industries, and only 200 kg of the raw material is converted into a usable product. About 600,000 tons of solid wastes annually are generated worldwide by leather industries [21]. An example of the types and quantities of solid wastes generated in leather processing based on one ton of raw hides/skin is given in **Table 1**.

The ways to disposals and valorizations for these wastes are defined by the chemical characteristics depending on the fact that the wastes are generated in either beamhouse or tanning and after tanning. This differentiation might be, namely, untanned wastes and tanned wastes accordingly.

Solid wastes generated from processing of raw hides/skin (1000 kg)	Quantity (kg)
Conservation salts	80
Hair	100
Raw trimmings	40
Lime sludges	60
Fleshings	120
Wet-blue trimmings	30
Chrome splittings	65
Chrome shavings	95
Buffing dusts	65
Crust trimmings	35
Dry sludge from common effluent treatment plants (CETPs)	125

Table 1.
Solid wastes from tannery [22].

2.1 Untanned solid wastes

Most of the solid wastes are generated in beamhouse, especially in fleshing operation. Fleshings are solid wastes generated during a mechanical process aiming at removing the flesh deposits or fats from the inner part of the skin [23]. Fleshings contain subcutaneous tissue, fat and flesh, which are composed of protein (5–7%), fat (4–18%), lime (2–6%), sulfide (2–4%), etc. [23].

Trimming is to cut out unwanted parts of processed hides/skin just after fleshing operation is completed. Trimmings are cut-outs from the operation and may be collected and shipped to glue manufactures or other by-product manufacturers or sent for disposal in a landfill [24].

Hides are generally subjected to mechanical operation called splitting to divide the hide into two or three layers horizontally. Splitting operation can also be applied at chromium tanning stage (wet-blue stage), which is called wet-blue splitting. Whether split is untanned and obtained after liming or tanned and obtained after tannage, it is a valuable part of a hide, which is a fibrous sheet, and hence it is in fact not a waste and more precisely it is a by-product.

The untanned solid wastes, mainly including leftovers from trimming of rawhide and surplus parts after liming and fleshing, are composed of large amount of collagen and grease. The chemical composition of these solid wastes varies depending on types and quality of the raw hides/skin and also process conditions. Fats and proteins are the main components of these wastes (10.5%). Moisture amounts might be up to 60%, meaning a high water content. The aforementioned solid wastes do not contain chromium compounds [25]. For sufficient usage of these protein-rich wastes, various kinds of methods and technologies have been proposed, focusing on the extraction of collagen/gelatin by using acid, alkali and enzyme hydrolysis and subsequent purification processes. Moreover, grease residue can also be used to extract oils and fats, which can be raw material for biofuel and leather fatliquor [26].

2.2 Chromium-tanned solid wastes

The chromium tanning is based on the cross-linkage of chromium ions with free carboxyl groups in the collagen. Chrome-tanned leather also called wet-blue leather are characterized by top handling quality, high hydrothermal stability, user-specific properties and versatility [27]. At the end of the chrome-tanning process, 60–75% of the chrome offer (Cr_2O_3) remains in the collagen structure. Additionally, small amounts of other chemicals and auxiliaries such as tensides, acids and bases (in the form of soluble "reaction salts") remain in the wet-blue leathers. The main environmental impact of tannery solid wastes is the oxidation of trivalent chromium into the hexavalent form, which is highly toxic and has carcinogenic and mutagenic effect. Leakages from chromium-containing wastes when they come to the agricultural lands cause ground water pollution and soil contamination. Water pollution affects aquatic animals, which are common sources of food, and soil contamination poses health effects through food chain and also poses a health hazard through inhalation of toxic dust, which can be inhaled by both people and livestock [28].

The solid wastes containing chromium namely tanned wastes are wet-blue shavings, wet-blue trimmings, buffing dusts, finished leather trimmings and wastewater treatment sludge [29]. Their chemical composition consists of fats and oils (3–6%) and mineral matters (15%). As chromium has been already used worldwide, they normally contain 3.5–4.5% of chromium as Cr_2O_3. Sludge from effluent treatment plants contains mainly water (up to 65%), organic substances (30%) and chromium (III) (around 2.5%) [25, 30].

Chrome shaving wastes are generated during the machine process of thickness adjustment of wet-blue leathers based on the required thickness. Shavings are mainly the scraps from the flesh side of leather, which are carried out by cutting unusable parts of leather and rags created during shaving operation [31]. Utilization or safe disposal of shavings continues to pose a serious challenge in many countries and is more critical because of their compositions. While processing one ton of raw hide, approximately 95–100 kg of wet-blue shavings are produced [32, 33]. Currently, a part of the chrome shavings is used in the manufacture of different types of areas such as leather board, collagen peptides, gelatin, animal feed and fertilizers. Unused portion of shavings is dumped in open areas around tanneries posing a serious environmental hazard [34].

2.3 Environmental and health impacts of leather solid wastes

The tannery solid wastes can cause severe problems associated with its organic load, inorganic matter, chromium, suspended solids, total organic and ammoniacal nitrogen, sulfide, and chloride, among others, depending on the chemical and mechanical processes applied to the raw hides/skin. Accumulation of these wastes leads to sludge problem and choking of treatment pipes and finally results in the reduction in efficiency of the treatment plant [35].

Leather industry is facing a lot of solid waste problem and many tanneries are closed for not meeting biological oxygen demand (BOD) and total dissolved solids (TDS) norms [4]. It is very important to analyze the nature of these wastes in order to assure a safe disposal or application of them. Salt, which is used to preserve hides or skin, discharges huge amount of pollution load in terms of total dissolved solids and chlorides and creates groundwater pollution [36]. Hair waste and lime sludge if discharged along with the effluents are likely to choke the drains. Trimmings, raw fleshings, limed fleshings and splitting waste can putrefy easily by producing noxious smells. Some of the biodegradable tannery solid wastes cause volatile organic compound emissions and, moreover, are sources of pathogenic bacteria [37].

Shaving dust contains environmentally unfriendly chemical called chromium, and when it is dumped in the environment, it can easily enter into the surface and ground, and this heavy metal pollutes the surface water by erosion and the underground water by leaching and erosion, leading to serious health problems to aquatic life in nearby rivers. As a result of this, pollution of surface and ground water results in shortage of drinking water for human beings and animals living at the downstream of the rivers [38].

According to Mu et al. [39], about 25% of tannery solid waste ends up as chromium-containing solid waste, which is more dangerous than other tannery solid waste. The waste generated from chrome-tanned leather is not biodegradable and toxic due to the chromium content [40]. Chromium-containing leather waste has been classified as one of the dangerous and hazardous waste if discharged into the environment without any pretreatment. Increased risks for a number of cancers such as lung cancer, testicular cancer, soft tissue sarcoma, pancreatic cancer and bladder cancer have been reported [41]. Chromium waste can also cause respiratory problems, a lower ability to fight disease, birth defects, infertility and tumor formation [42]. Chromium-containing solid waste percolates to the ground and causes ground water pollution and soil contamination. Water pollution affects aquatic animals that are common sources of food, and contamination of soil poses health effects through food chain and also poses a health hazard through inhalation of toxic dust, which can be inhaled by both people and livestock. It can damage the gills of fish; it can alter genetic materials and cause cancer [43]. Moreover, thermal incineration of these wastes is associated with serious air pollution problems due to emission of toxic hexavalent chromium (Cr+6), halogenated organic compounds, aromatic hydrocarbons, etc. into the environment [42].

3. Collagen peptides

The word collagen is derived from the Greek word "kola," which means gum, and "gen," which means producing. It is a fibrous structural protein present in extracellular matrix and connective tissue of animals [44]. Collagen is the most prevalent protein comprising approximately 30% of the total protein of animal and human bodies and is found primarily in connective tissues including animal hairs, bones, cartilages, tendons and blood vessels [45, 46].

There are many types of collagens, which are from collagen I to collagen XIX. Animal skin or hide contains collagen type I, mostly with approximately 90% of its dry weight [47]. Collagen precursors are synthesized in the endoplasmic reticulum of cells and transported to the Golgi apparatus in order to secrete into the extracellular spaces, and maturation of collagen can occur [47, 48].

Collagen polypeptide chains and cross-linkages can be broken down by hydrolytic processes and decomposition yielding in different subunits and fragments [49]. In other words, gelatin is produced by partial denaturation of collagen in triple helical structure. Gelatin and collagen peptides are new forms yielded by hydrolysis of native collagen with lower molecular weight fragments than original structure and including a wide range of subcategories having differentiated functionalities [50]. Native collagen exhibits superior and distinct properties from collagen peptides such as higher enthalpy, greater network structure of fibrils, basic isoelectric point and high resistance to protease hydrolysis [51]. The native triple helices and fibril networks in the native collagen are more rigid and firmer than gelatin and collagen peptides [52].

3.1 Structural and chemical characteristics of native collagen

All proteins are composed of linear chains of amino acids attached together by peptide bonds, thereby making oligomers, which are the primary structure of the protein. Being arranged into sequences of different amino acids gives way to fold up the chains into a functional protein. Intermolecular and intramolecular bonds in the structure can enhance the folding of the peptide chains. This folding induces the weak forces such as hydrogen bonds and electrostatic, hydrophobic and van der Waals interactions [53]. Considering the reactivity of collagen molecule, the peptide bond itself is prioritized, capable of participating in hydrogen bonds with both hydrogen-bond donor and acceptor groups. For example, a carbonyl group in any protein has two lone pairs of electrons having the capability of accepting hydrogen bonds. In addition, the electronegative nitrogen induces a partial positive charge on its attached hydrogen, allowing the hydrogen to function as a hydrogen-bond donor [54]. Hydrogen bonding between peptide bonds is the basis of protein secondary structural formation, namely, helices, yielding in pleats and turns in the structure [55]. The side chains of amino acids are capable of a variety of interactions including hydrogen bonds, ionic bonding, hydrophobic interactions, van der Waals interactions and disulfide bonds. These interactions and secondary structural elements are responsible for taking a shape of the tertiary structure of proteins, their actual three-dimensional shape [56]. Due to the interactions between side chains of various amino acids, the protein molecule will bend and twist so as to gain individual stability or lower energy state.

Proteins included in the number and arrangements of subunits to give functionality are referred to as quaternary structure. The proteins comprising individual subunits may be identical, or they may be different. Like the secondary and tertiary structures, the quaternary structure of a protein is determined by its primary structure [57].

Collagens are trimeric molecules made up of three polypeptide chains, which contain the sequence repeat of (Gly-X-Y)n, X being frequently proline and Y hydroxyproline. These repeats allow the formation of a triple helix based on three polypeptide chains bound to each other by hydrogen bonding, which is the characteristic feature of the collagen [58]. The side chains of each X and Y residue are at the surface of triple helix, giving the collagen molecule a significant capacity for lateral interactions with other molecules of extracellular matrix and resulting in the formation of various supramolecular assemblies [59, 60]. An interchain hydrogen bonding between glycine and amide group in an adjacent chain is a key factor in stabilizing the collagen triple helix [61].

Collagen protein is more hydrophilic than lyophilic moieties due to the chemical nature of numerous amino acids present in its structure [62]. It has a highly complex structure and interacts with each other at the molecular level to form broader systems with distinctive properties [63]. The chemical structure of collagen type I is shown in **Figure 1**.

Basic properties create characteristic structure of collagen fibril helicoidal structure is; fibril diameters ranging from 10 to 500 nm [64], average molecular weight of 285,000 Da [65] and glycine ratio of 1/3 in the polypeptide chain consisting of 1400 amino acids [66].

Having been readily recognized in tissues with commonly white and opaque colors, collagen fibers area considered as viscoelastic materials having high tensile strength and low extensibility. The tensile strength of collagen depends on the formation of covalent intermolecular cross-links between the individual protein subunits [67].

The collagen family is highly complex and shows a remarkable diversity in molecular and supramolecular organization, tissue distribution and function. Collagen types are classified in several subfamilies according to sequence homologies, similarities in their structural organization and supramolecular assembly. The availability of 27 collagen types was reported and they are classified by their size, function and amino acid distribution that differ considerably in their

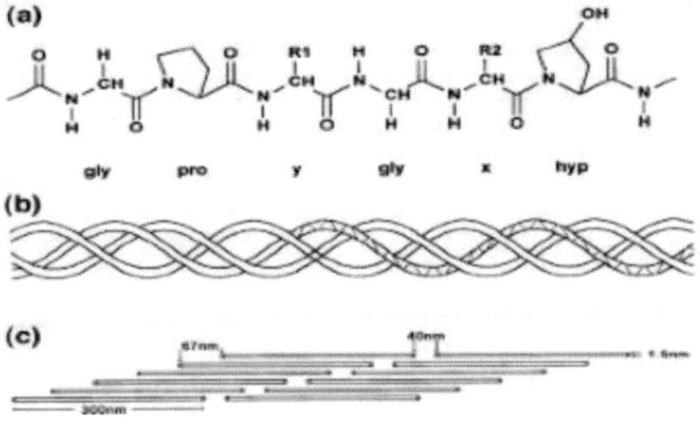

Figure 1.
Chemical structure of collagen type I. (a) Primary amino acid sequence, (b) secondary left-handed helix and tertiary right-handed triple-helix structure, and (c) staggered quaternary structure [63].

molecular structure [68]. The individual members are numbered with roman numerals. The family is subdivided into different classes: the fibrillar collagens (types I, II, III, V, XI, XXIV and XXVII), basement membrane collagens (type IV), fibril-associated collagens with interrupted triple helices (FACIT collagens, types IX, XII, XIV, XVI, XIX, XX and XXI), short chain collagens (types VIII and X), anchoring fibril collagen (type VII), multiplexins (types XV and XVIII), membrane-associated collagens with interrupted triple helices (MACIT collagens, types XIII, XVII, XXIII and XXV) and collagen type VI. The types indicated by an asterisk are heterotrimers, consisting of two or three different polypeptide chains. Type IV collagens contain six different polypeptide chains that form at least three distinct molecules and type V collagens contain three polypeptide chains in probably three molecules [68].

Each collagen type has its own specific amino acid composition and performs a distinctive role in tissues. Types I, II and III are of the most abundant collagens, which are responsible for tissue strength, elasticity and water retention capacity [69]. Type I collagen is the main structural component of extracellular matrix. It consists of one α2 chain and two α1 chains, which are encoded on chromosome 7 and 17 in humans [70]. Generally, type 1 collagen is the most commonly used in industrial scale especially in tissue repair and replacement, and they are intensive in skin, tendon, bone, cornea, dentin, fibrocartilage, large vessels, intestine, uterus, dermis, cornea and connective tissue [71]. It has outstanding mechanical properties and is present in virtually every extracellular tissue with mechanical function. In tendons and ligaments, collagen transmits the force from muscles to bones and stores elastic energy. Smooth walking would not be possible without these properties. Collagen also represents most of the organic matrix of bones and tooth dentin and confers them their fracture resistance. It is a major constituent of skin and blood vessels and is even present in muscles, which could not function without a collagen-rich matrix around the contractile cells. A slightly different type of collagen type II is a critical component of a tissue as soft as articular cartilage. The function of collagen is not only mechanical. In the cornea of the eye, for example, the ordering of collagen fibrils confers transparency in addition to mechanical stability [69]. Type II collagen is prevalent in hyaline cartilage, vitreous, nucleus pulposus, notochord and intervertebral disc. It provides biomarkers for osteoarthritis. Type III collagen is present in fetal dermis and epidermis, veins, uterus, synovium, connective tissue around muscles and also in small quantities in areas where type I collagen is present. Type III collagen is functional of fibrillogenesis of collagen I and for normal cardiovascular development [72].

3.2 Industrial sources of collagen peptides

The major sources of collagen for fabrication are bovine and porcine species, where collagen was extracted from the hides and skin and also bones of pigs and cows. Bovine hides, a by-product of meat production, are one of the major industrial sources of collagen [49]. The bovine hide is composed of approximately 30% protein, and the inner corium layer of the hide is rich in collagen. This collagen has a high denaturation temperature in comparison to collagen from other sources. Bovine hide is practiced upon in different development stages such as bovine dermis used for tendon regeneration, and skin and wound healing (in the form of collagen matrix); neonatal bovine dermis is used for hernia repair, plastic and reconstructive surgery [73].

Starting from the 1930s, the most significant raw material for large-scale industrial gelatin production is porcine skin [74]. The skin and bones of pigs are utilized as a collagen sources due to some advantages. Since porcine collagen is almost similar to human collagen, it does not cause much allergic response when used in health applications. But just like the bovine source, the zoonotic diseases poses a risk of contamination and pigs are proscribed due to religious reasons [60]. Halal certification of collagen derivatives is considered to be of main importance because of beliefs and it depends on the origin of raw materials used in its manufacture and traceability from the sources until product chain. Muslims and Jew people demand Halal-certified products for their needs, which is not prohibited and obtained by entirely traceable product chains. Nonspecific collagen is highly suspected of containing porcine elements and very strongly discouraged for use by the Muslims [75]. Nonetheless, adult porcine dermis and small intestinal mucosa are used for tendon regeneration, hernia repair, skin and wound healing, and plastic and reconstructive surgery [76].

There are some other sources of gelatin, somehow industrially applicable or not. Throughout the decade, huge numbers of fish species were investigated as alternatives to the source of collagen. Bones, skin, fins and scales of fresh or salt water fishes are mainly used for collagen procurement and gelatin extraction having different chemical composition. This in turn helps to reduce environmental pollution as considerable amount of wastes occurs during fish processing [77]. Collagen studies from marine origin are carried by on marine vertebrates and invertebrates [78, 79]. Marine sources are from some marine species such as fishes, starfish, jellyfish, sponges, sea urchin, octopus, squid, cuttlefish, sea anemone and prawn [80–82]. Some of the raw material sources of collagen peptides are given in **Figure 2**.

Collagen peptides can also be produced for research purposes in small quantities from other animal body parts such as eggshells, rat-tail tendons, frog skin, kangaroo tails, chicken and duck feet, sheepskin, poultry animal skin, feet, bones and many more [46, 83, 84].

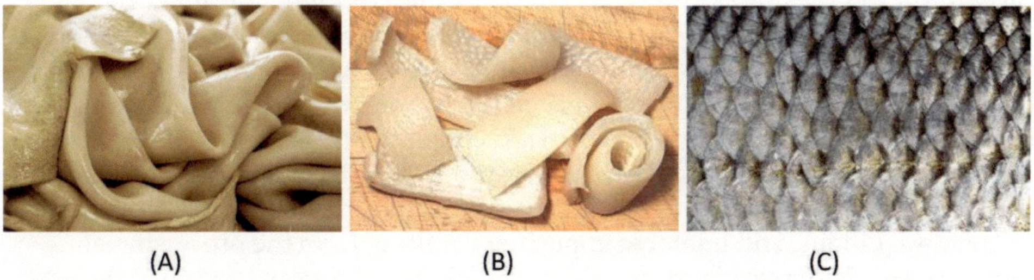

Figure 2.
Main raw material sources of collagen peptides. (A) Bovine split, (B) pig skin, and (C) fish skin.

4. Collagen peptide processing

The collagenic substances, which are involved in multiple collagen units (not subunits) in the quaternary structure and arguably misdefine the tertiary structure also, are normally processable structures. As known very well, leather manufacturing can convert low-value raw materials, which unless untanned and disposed of have detrimental effect to the environment, into valuable final products, and collagen is one of the most substantial structural protein, economically and biologically renewable material for processing. The wastes and by-products of leather processing are discarded parts and effluents from many steps, which are still valuable due to their composition. Lime splits and scraps, as ideal substances, go for gelatin and collagen peptide production.

In the production of industrial-scale collagen peptides, different animal's skin and bones that are easily available and contains collagen protein in high proportion are being used. Collagen peptide preparation steps are dependent on final products' properties. For the first step in general practice, acid and alkali extraction methods are used to remove noncollagenous components [85].

Enzymatic and chemical hydrolysis can be used in the extraction of collagen. Being affordable, chemical hydrolysis is the most commonly used method in industrial practices. Enzymatic hydrolysis is fast and produce waste in minimal amounts, but they are more expensive to carry out [86].

Collagen peptides can be produced by sensitive enzymatic reactions according to the desired molecular weights from collagen-rich raw materials by using protease enzymes. Depending on enzyme types and hydrolyzation conditions, final products can further differ with regard to molecular weight distribution [87]. The production processes could be optimized to obtain different peptides with different functionalities. In the structural level, the cleavage of triple helix is emerged and the collagen molecule is partially broken up. Long chains are hydrolyzed to form shorter chains and further hydrolysis leads to short peptides, some of which are bioactive with body-stimulating functions [88].

Chemical methods of collagen hydrolysis are carried out by means of strong acidic and highly alkaline conditions. Acid and alkaline hydrolysis methods are cost-effective and operation is simple. They have short hydrolysis time and are applicable to industrial processes [89]. However, the uses of strong acids or strong alkaline chemicals make the hydrolysis process environmentally unacceptable [90]. During the acidic treatment, the raw material is exposed to acid for a certain period of time. As this process occurs at a controlled temperature, the structure of the skin swells to twice or thrice more than its initial volume. Both organic acids such as acetic and citric acids and inorganic acids such as hydrochloric acid can be used during acid treatment; however, organic acids are more efficient for the purpose. Acidic treatment results unraveled the structural unity and the cleavage of the noncovalent inter- and intramolecular bonds. Materials with less intertwined collagen fibers such as fish and porcine skin are the preferred choice for the acidic process [85]. For the alkaline process, the raw materials are treated in basic solutions for a duration of a few days to weeks. The most commonly used process is through aqueous sodium hydroxide and calcium hydroxide solutions. However, other basic solutions can also be used in this alkaline process. This process entails the treatment of hard or thick substance that needs very aggressive penetration by the basic solutions [91].

To meet the technical needs of the different sectors, purification stage ensures the removal of ionic and nonionic impurities resulting from the processing of raw materials. Different filtration and purification systems can be used at this stage depending on the final product needs [92]. The purified and demineralized gelatin solution consists of over 95% water. This water has to be almost completely removed. Only dried gelatin with its normal residual water content of 10–12% has an unlimited shelf life from the microbiological point of view. In addition, dilute gelatin solutions can neither be stored nor transported easily. In the next production step, the highly concentrated and filtered gelatin solutions are sterilized. For this step, both indirect sterilization via plate heat exchangers and direct steam sterilization are used. Both methods are microbiologically safe to a very high degree [93]. After sterilization, the prepared material needs to be dried to final form. There are different drying methods used in the production of collagen peptides. Spray drying is the most commonly used method and widely used in the production of small molecular weight peptides [94].

5. Industrial applications

5.1 Food

Collagen peptides have shown to be an important ingredient in the food and beverage industries worldwide [95]. It has been used for a long time in foods globally, such as in the United States, China, Japan and many countries in Europe. Approved as Generally Recognized As Safe (GRAS), the safety of collagen peptides has been affirmed by the Food and Drug Administration (FDA) and Center for Food Safety and Applied Nutrition (CFSAN) [96]. It has been applied as protein dietary supplements, carriers in the meat processing, edible film and coatings of products and food additive to improve product's functionality [97]. In addition, collagen may boost the health and nutritional value of the products relying on its inimitable properties on human bodies [75].

The source of the raw material and the degree of processing determine the properties of the collagen peptides like gelatin, which have several different applications in the food industry [98]. The major quality parameters are their higher gel strength and suitable melting and gelling temperatures for the food industry that uses them as an additive. Due to the fact that porcine and bovine gelatins are less preferred due to religious preferences, safety concerns and economic considerations, using fish skin or bones to obtain gelatin has become popular in recent years [99]. Thanks to its many unique properties, the numerous applications of gelatin include its usage as a thickener, stabilizer, setting agent, clarifying agent, water-retaining agent and adhesive in a wide range of foods, pharmaceuticals and household products. In the food industry, gelatin can be utilized in a wide range of confectioneries, beverages, snacks, desserts and meat products [100]. Gelatin is used as an additive to improve elasticity, consistency and stability of foods like desserts, candies, bakery products, jellied meats, ice cream and dairy products. Gelatin is also used as stabilizer to modify the structure of the food products. It is added to yogurt to reduce syneresis and increase firmness [100]. In addition, type A gelatin that is isolated with acid treatment with gel strength as 70–90 g, which is relatively low, is used to fine wines and juices. Type B gelatin is processed with an alkali treatment with gel strength as 125–250 g and is used in confectionery products [101]. Collagen peptides have also been reported to have antioxidant and antimicrobial activity [102]. However, the relationship between peptide characteristics and antimicrobial activity has not been clearly demonstrated.

5.2 Cosmetic

Collagen can be used in cosmetics due to its biodegradability, availability and biocompatibility properties for different purposes such as in dermal fillers, skin substitutes or scaffolding, wound repairs and facial products [103].

The formation of unwanted wrinkles in the body with aging is related to the damage of the fibers in the skin. In the researches about aging, it has been determined that collagen hydrolysates contribute greatly to the repair of these fibers [104]. The introduction of collagen hydrolysate into the body ensures the stimulation of collagen formation that enables the recovery and improved tissue appearance [105]. Hence, the cosmetic industry reclaims some functionalities of its products by incorporating this biomolecule.

Collagen peptide has been known to be used in cosmetic formulations for reasons such as protecting the structure and the function of the skin, enhancing its appearance and preventing premature aging [106].

Collagen peptide is prepared in the form of liquid ampoules, powder mixes or tablets in the food and cosmetic industries. It has a regenerative effect on skin wrinkles and other signs of skin aging: collagen helps the skin remain soft and pliant and improves the hydration of the epidermis [107]. Many studies have shown that collagen sleek thin lines and can prevent the development of deeper wrinkles and grooves. Collagen is not only effective for the skin on the face but also stimulates the fiber structure of the body to repair and reduce cellulite tissue [108].

Collagen hydrolysates have also shown bioactivities such as antioxidant properties, antihypertensive activity, lipid-lowering activity, as well as reparative properties in damaged skin [109]. Moreover, it has been also observed that collagen provides the building block for elastin and collagen formation and acts as ligands in fibroblast cells to stimulate hyaluronic acid [110].

5.3 Health

Collagen is the most abundant and ubiquitous protein in the body regarded as one of the most useful biomaterials. The excellent biocompatibility and safety due to its biological characteristics made collagen the primary resource in medical applications. It has various applications in some departments such as cardiology (heart valve), dermatology (for skin replacement, augmentation of soft tissue, skin tissue engineering and artificial skin dermis), surgery (as hemostatic agent, wound repair and dressing, nerve repair and blood vessel prostheses), orthopedy (tendon, bone and ligament repair and cartilage reconstruction), ophthalmology (corneal grafts and contact lenses), urology (hemodialysis and sphincter repair) and vascular surgery (vascular graft and vessel replacement) [111].

Collagen type I is considered to be the most valuable material for tissue engineering due to its high biocompatibility and immunogenicity. It is used as the basic matrix for cell culture [73, 112]. Biomaterials based on collagen are widely used in tissue engineering such as injectable matrices and scaffolds intended for bone regeneration [73, 113]. Moreover, collagen-based eye implants are preferred for the treatment of ophthalmic disorders. Such type of collagen-based implant preparation has shown considerable applicability because it provides stable and reasonable control over the postoperative complications such as intraocular pressure [114]. Collagen-based matrices find their use as corneal transplant and as temporary patches to repair perforations in case of emergencies [115].

Collagen is used in pharmaceutical industries for different functionalities as hard and soft dry capsules, microparticles, injectable dispersions, shields in ophthalmology sponges and drug delivery system. Its application in the pharmaceutical as well as biomedical field is due to its characteristics such as weak antigenicity, immunogenicity, biodegradability and biocompatibility [116].

As a collagen peptide, gelatin is the most important material for the production of hard and soft capsules as well as film-coated and effervescent tablets. Manufacturers take into account its adhesive, gelling and film-building properties.

Orally administered medicines and dietary supplements in particular are protected by gelatin-containing capsules or tablets from light, moisture and oxygen and given a long shelf life [107, 117, 118]. Gelatin is also used as a raw material in many field of health industry as is the case with the manufacture of blood substitute [119]. These products prevent hypovolemic shock by stopping bleeding in the wound-occurred area. As local hemostatic agents, collagen sponges and films have long been used in the surgical field (e.g., in oral cavity and ophthalmological surgery, urology or gynecology) and for the treatment of wounds in dental surgery. The structural composition of the collagen material enables the absorption of large amounts of blood and makes it possible for new tissue to grow into the sponges. Since it only takes a few days for the body to completely resorb the sponges or films, they can be left in the wound without any negative effects [107, 120, 121].

5.4 Sportive nutrition

Collagen peptides are ideal supply due to their numerous beneficial health effects for modern sportsperson nutrition as high-energy supplement to maximize muscle protein anabolism [122]. They are neutral in flavor, which means that they do not leave a bitter aftertaste that has to be masked in the final product, that is, through sugar or artificial sweeteners, as is often the case with soy, whey or other protein [107, 123]. Collagen peptides have been scientifically tested and have no undesirable side effects, and there is no evidence to elicit allergic reactions. It emulsifies foams and improves the shelf life of products [107, 124].

The more protein a body expends through physical exertion, the greater its needs for an external source, for example in the form of special dietary supplements such as protein shakes, energy bars, protein snacks or sports drinks. Several studies in the past few decades have reported that protein hydrolysates from various food sources, in addition to their nutritional properties, exhibited various biological functions including hypotensive activity, anticoagulant, cholesterol-lowering ability and hypoglycemic effect [125]. Consumption of hydrolyzed collagen increases collagen synthesis and decreases knee pain while standing and walking [126]. Shaw et al. [127] tested the role of gelatin consumption in collagen synthesis. In the study, double-blinded, placebo-controlled and crossover-designed research, subjected to whom consumed 15 g of gelatin showed double-fold collagen synthesis, measured through serum propeptide levels. From the results, it was observed that consuming hydrolyzed collagen might increase collagen synthesis and potentially decrease injury rate in athletes. Studies have also shown that products fortified with collagen peptide can promote joint health, bone synthesis and antisport fatigue ability [128].

5.5 Agriculture and animal feed

5.5.1 Fertilizer

Leather processing wastes like shavings that cause environmental pollution are opulent sources of novel and valuable biomolecule "collagen" [129]. Industry has been generally oriented on the recovery of collagen from leather waste, but the remaining waste also can be used for agricultural purposes. Collagen-based fertilizer products highly are demanded in agriculture industry because of being high amino acid and organic carbon source and nitrogen content [130].

The collagen hydrolysates obtained from leather wastes are being utilized as biofertilizer. Several plants can also take up and absorb amino acids as an example of biostimulants; these amino acids are sometimes better nitrogen sources than ammonia or nitrates [131]. Collagen peptides are recovered and channeled as an organic nitrogenous fertilizer to increase the yield of the crop [132]. Both plants and animal organisms can more easily absorb microelements like iron, copper, zinc, calcium, magnesium and manganese chelated with hydrolyzed collagen. The use of collagen hydrolysates in combination with potassium polyphosphates increases agricultural production by increasing the absorption of phosphorus and potassium [133].

Collagen hydrolysates obtained by chemical and chemical-enzymatic processes under moderate reaction conditions were used in a study for preparation of foliar fertilizers [134]. Hydrolysates of chromium-tanned leather shavings were used in a study as nitrogen source for growth of common bean plants and banana cultivation [135]. De Oliveira et al. [136] have studied the use of leather wastes after extraction as a nitrogen source to elephant grass. The chrome shaving wastes can also be hydrolyzed in an autoclave (150°C). The obtained product contains moisture content (7–10%), total nitrogen (10–11%), organic carbon (40%) and chromium (III) (2.5–3%). By blending with other additive components, the product can be sold as a fertilizer [133].

Both gelatin and collagen hydrolysates have positive effect on the growth of plants when applied as fertilizer. The crop yield is comparable with those obtained by using inorganic fertilizers but with a significantly high value in view of the low nitrate content, which is 20 times less. Besides, organic fertilizer improves the soil quality unlike the inorganic ones [137].

5.5.2 Animal feed

Collagen peptide due to its organic compounds such as fats, proteins and minerals plays an important role in the preparation of highly valuable animal feed [84]. Fat, protein and mineral products are in especially high demand in the animal feed industry because pure fats are excellent sources of energy and collagen is of importance for the healthy growth of animals [138].

As a collagen peptide, which is recovered from leather solid wastes, gelatin is primarily added to animal feed based on its hydrophilic properties. Its jelly-like consistency holds feed together, making it transportable and extends its shelf life [107, 139]. When animal feed is enriched with vitamins, the gelatin coatings also protect these from light and oxygen. A positive side effect of adding gelatin to feed ensures that the fur of animal remains wonderfully glossy [107, 140].

There are also many fields for collagen peptides and gelatin usage. They can be used for photography and X-ray films and inkjet applications [141], industrial paper production [93], leather board [142], glue manufacture [143], feedstock for biodiesel production [144], leather tanning and retanning agent [145] and many more specific applications, etc.

6. Conclusion

The tanning industry is one of the oldest industries in the world and recently its pollution load onto environment has become seriously threatening for transferring the potential to next generations. It produces a significant amount of solid wastes and effluents. It is a well-known fact that removing undesired substances out of the structure in leather processing produces effluents; that is highlighted agenda which needs to be overcome and as per the composition those are able to handle for recover and reuse through the current technology.

Revaluation of leather solid wastes is one of the promising waste management strategies that provides raw materials to another industry such as food, agriculture, cosmetic, health, etc. This method may offer a solution for utilization of huge volume of leather solid wastes, which are often dumped in open landfills. Commercial benefits of the system should be linked with both the value of the products and the disposal cost of solid wastes. In the economical point of view, feasibility should be based on converting them into value-added products instead of making a deposit for disposal.

Collagen peptides obtained from hides and their by-products have been practiced as healthful stuffs in many areas of our modern life. As the awareness of their technological value increases as time passes, this value-added material is considered to have higher interest with the usage in various fields. Analogous to collagen peptides, collagen hydrolysates and gelatins, emerged from this precious protein are involved in either partly or total denaturation. It is the process defined by disintegration of intra- and intermolecular bonds that keeps together the chains composed of amino acids in the conformation; thereby, a typical protein is formed. The discovery of benefits of collagen derivatives for health and their usage as additives has a long history and is dated back to some 8000 years ago. Today, its usage enlarges over many industries and applications includes in food, health, chemical, body care and agricultural etc. According to the molecular weight and properties, the usage of collagen derivatives increases as a gel or colloidal solution, their benefits are multiplied and this bio-based material supplies many valorization possibilities. As per the source and properties, they are bioavailable products that are digested and absorbed by human and animal body quickly and even by plants and are also easy to use in any industrial applications and processes.

In spite of the tremendous development in technology and sciences, there are still challenges ahead to better understand the collagen types and sources, structure and properties, gelatin processes and product characteristics. It seems that in the future the researches on bio-based materials as well as the efforts for their commercialization will continue intensively in a wider range of products.

Conflict of interest

The authors declare that they have no conflict of interest.

Author details

Ali Yorgancioglu[2*], Bahri Başaran[1] and Aykut Sancakli[2]

1 Faculty of Engineering, Department of Leather Engineering, Ege University, Bornova, İzmir, Türkiye

2 Kazlicesme R&D Center and Testing Laboratories, Istanbul, Türkiye

*Address all correspondence to: ali.yorgancioglu@ege.edu.tr

References

[1] Irshad A, Sharma BD. Slaughter house by-product utilization for sustainable meat industry: A review. ournal of Animal Production Advances. 2015;**5**:681-696. DOI: 10.5455/japa.20150626043918

[2] Thomson Kite M. The nature and properties of leather. In: Conservation of Leather and Related Materials. Jordan Hill, Oxford: Elsevier Linacre House; 2006. p. 1

[3] Ork N, Ozgunay H, Mutlu M, Ondogan Z. Manufacture of leather skirt from garment leathers tanned with various tanning materials and evaluation of visual properties. Tekstil ve Konfeksiyon. 2017;**27**:91-97

[4] Wang Y, Zeng Y, Zhou J, Zhang W, Liao X, Shi B. An integrated cleaner beamhouse process for minimization of nitrogen pollution in leather manufacture. Journal of Cleaner Production. 2016;**112**:2-8. DOI: 10.1016/j.jclepro.2015.07.060

[5] Şengil IA, Kulaç S, Özacar M. Treatment of tannery liming drum wastewater by electrocoagulation. Journal of Hazardous Materials. 2009;**167**:940-946. DOI: 10.1016/j.jhazmat.2009.01.099

[6] Mendoza-Roca JA, Galiana-Aleixandre MV, Lora-García J, Bes-Piá A. Purification of tannery effluents by ultrafiltration in view of permeate reuse. Separation and Purification Technology. 2010;**70**:296-301. DOI: 10.1016/j.seppur.2009.10.010

[7] Paschoal FMM, Anderson MA, Zanoni MVB. Simultaneous removal of chromium and leather dye from simulated tannery effluent by photoelectrochemistry. Journal of Hazardous Materials. 2009;**166**:531-537. DOI: 10.1016/j.jhazmat.2008.11.058

[8] Fathima NN, Balaraman M, Rao JR, Nair BU. Effect of zirconium(IV) complexes on the thermal and enzymatic stability of type I collagen. Journal of Inorganic Biochemistry. 2003;**95**:47-54. DOI: 10.1016/S0162-0134(03)00071-0

[9] Covington AD. Quo vadit chromium? The future direction of tannage. Journal of the American Leather Chemists Association. 2008;**103**:7-23

[10] Covington AD. Tanning Chemistry, The Science of Leather. Northampton, UK: The University of Northampton; 2009

[11] Mohamed OA, Moustafa AB, Mehawed MA, El-Sayed NH. Styrene and butyl methacrylate copolymers and their application in leather finishing. Journal of Applied Polymer Science. 2009;**111**:1488-1495. DOI: 10.1002/app.29022

[12] Ozgunay H, Colak S, Mutlu MM, Akyuz F. Characterization of leather industry wastes. Polish Journal of Environmental Studies. 2007;**16**:867-873

[13] Mella B, Benvenuti J, Oliveira RF, Gutterres M. Preparation and characterization of activated carbon produced from tannery solid waste applied for tannery wastewater treatment. Environmental Science and Pollution Research. 2019;**26**:1-7. DOI: 10.1007/s11356-019-04161-x

[14] Aceves MB, Velasquez RO, Vazquez RR. Effects of $Cr3+$, $Cr6+$ and tannery sludge on C and N mineralization and microbial activity in semi-arid soils. Journal of Hazardous Materials. 2007;**143**:522-531. DOI: 10.1016/j.jhazmat.2006.09.095

[15] Cavalcante DGSM, Gomes AS, Santos RJ, Kerche-Silva LE, Danna CS,

Yoshihara E, et al. Composites produced from natural rubber and chrome tanned leather wastes: Evaluation of their in vitro toxicological effects for application in footwear and textile industries. Journal of Polymers and the Environment. 2018;**26**:980-988. DOI: 10.1007/s10924-017-1002-9

[16] Sastry TP, Sehgal PK, Ramasami T. Value added eco-friendly products from tannery solid wastes. Journal of Environmental Science & Engineering. 2005;**47**:250-255

[17] Song H, Li B. Beneficial effects of collagen hydrolysate: A review on recent developments. Biomedical Journal of Scientific & Technical Research (BJSTR). 2017;**1**:1-4

[18] Ola AM, Taha GM, Elsayed NH. Technical gelatine from chrome tanned shavings. International Journal of Environment and Waste Management. 2013;**11**:335-349

[19] Basaran B, Yorgancioglu A, Onem E, Bitlisli BO. Performance assessment of green practices in liming with reductive potential chemicals for environmental sustainability. Revista de Pielarie Incaltaminte / Leather and Footwear Journal. 2019;**3**:177-184

[20] COTANCE. Social and Environmental Report. The European Leather Industry; 2012

[21] Onukak IE, Mohammed-Dabo IA, Alewo A, Okoduwa SIR, Fasanya O. Production and characterization of biomass briquettes from tannery solid waste. Recycling. 2017;**2**:17. DOI: 10.3390/recycling2040017

[22] Rao JR, Thanikaivelan P, Sreeram KJ, Nair BU. Tanning studies with basic chromium sulfate prepared using chrome shavings as a reductant: A call for 'Wealth from Waste' approach to the tanning industry. Journal of the American Leather Chemists Association. 2004;**99**:170-176

[23] Abul HM, Nur-A-Tomal M, Biplab KM. Solid waste generation during fleshing operation from tannery and its environmental impact: Bangladesh perspective. In: 2nd International Conference on Advances in Civil Engineering. Chittagong, Bangladesh: (ICACE-2014), At CUET; 2014

[24] Masilamani D, Madhan B, Shanmugam G, Saravanan P, Narayan B. Extraction of collagen from raw trimming wastes of tannery: A waste to wealth approach. Journal of Cleaner Production. 2016;**113**:338-344. DOI: 10.1016/j.jclepro.2015.11.087

[25] Kanagaraj J, Velappan KC, Babu NKC, Sadulla S. Solid wastes generation in the leather industry and its utilization for cleaner environment: A review. Journal of Scientific and Industrial Research. 2006;**65**:541-548

[26] Li Y, Guo R, Lu W, Zhu D. Research progress on resource utilization of leather solid waste. Journal of Leather Science and Engineering. 2019;**1**:1-17. DOI: 10.1186/s42825-019-0008-6

[27] Onem E, Yorgancioglu A, Karavana HA, Yılmaz O. Comparison of different tanning agents on the stabilization of collagen via differential scanning calorimetry. Journal of Thermal Analysis and Calorimetry. 2017;**129**:615-622. DOI: 10.1007/s10973-017-6175-x

[28] Sundar VJ, Raghavarao J, Muralidharan C, Mandal AB. Recovery and utilization of chromium-tanned Proteinous wastes of leather making: A review. Critical Reviews in Environmental Science and Technology. 2011;**41**:2048-2075. DOI: 10.1080/10643389.2010.497434

[29] Nigam H, Das M, Chauhan S, Pandey P, Swati P, Yadav M, et al. Effect of chromium generated by solid waste of tannery and microbial degradation of chromium to reduce its toxicity. Advances in Applied Science Research. 2015;**6**:129-136

[30] Ahamed MIN, Kashif PM. Safety disposal of tannery effluent sludge: Challenges to researchers-a review. International Journal of Pharmaceutical Sciences and Research. 2014;**5**:733-736

[31] Fathima N, Rao JR, Nair BU. Tannery solid waste to treat toxic liquid wastes: A new holistic paradigm. Environmental Engineering Science. 2012;**29**:363-372. DOI: 10.1089/ees.2010.0445

[32] Pati A, Chaudhary R, Subramani S. A review on management of chrome-tanned leather shavings: A holistic paradigm to combat the environmental issues. Environmental Science and Pollution Research. 2014;**21**:11266-11282. DOI: 10.1007/s11356-014-3055-9

[33] Saikia P, Goswami T, Dutta D, Dutta NK, Sengupta P, Neog D. Development of a flexible composite from leather industry waste and evaluation of their physico-chemical properties. Clean Technologies and Environmental Policy. 2017;**19**:2171-2178. DOI: 10.1007/s10098-017-1396-z

[34] Dwivedi SP, Dixit A, Bajaj R. Development of bio-composite material by utilizing chrome containing leather waste with Al2O3 ceramic particles. Materials Research Express. 2019;**6**. DOI: 10.1088/2053-1591/ab3f8e

[35] Sathish M, Madhan B, Rao RJ. Leather solid waste: An eco-benign raw material for leather chemical preparation – A circular economy example. Waste Management. 2019;**87**:357-367. DOI: 10.1016/j.wasman.2019.02.026

[36] Morera JM, Bacardit A, Ollé L, Bartolí E, Borràs MD. Minimization of the environmental impact of chrome tanning: A new process with high chrome exhaustion. Chemosphere. 2007;**69**:1728-1733. DOI: 10.1016/j.chemosphere.2007.05.086

[37] Kanagaraj J, Senthilvelan T, Panda RC, Kavitha S. Eco-friendly waste management strategies for greener environment towards sustainable development in leather industry: A comprehensive review. Journal of Cleaner Production. 2015;**89**:1-17. DOI: 10.1016/j.jclepro.2014.11.013

[38] Lakrafli H, Tahiri S, Albizane A, El Otmani ME. Effect of wet blue chrome shaving and buffing dust of leather industry on the thermal conductivity of cement and plaster based materials. Construction and Building Materials. 2012;**30**:590-596. DOI: 10.1016/j.conbuildmat.2011.12.041

[39] Mu C, Lin W, Zhang M, Zhu QS. Towards zero discharge of chromium containing leather waste through improved alkali hydrolysis. Waste Management. 2003;**23**:835-843. DOI: 10.1016/S0956-053X(03)00040-0

[40] Tahiri S, Bouhria M, Albizane A, Messaoudi A, Azzi M, Alami SY, et al. Extraction of proteins from chrome shavings with sodium hydroxide and reuse of chromium in the tanning process. Journal of the American Leather Chemists Association. 2004;**99**:16-25

[41] Sethuraman C, Srinivas K, Sekaran G. Conversion of hazardous leather solid waste into fuels and products C2013 Ieee global humanitarian technology conference: South Asia satellite. GHTC SAS. 2013:274-279. DOI: 10.1109/GHTC-SAS.2013.6629930

[42] Shanker AK, Venkateswarlu B. Chromium: Environmental pollution,

health effects and mode of action. In: Jerome ON, editor. Encyclopedia of Environmental Health. Vol. 65. Burlington: Elsevier; 2011. pp. 650-659. DOI: 10.1016/B978-0-444-52272-6.00390-1

[43] Harris J, McCartor A. The world's worst toxic pollution problems report. In: The Top Ten of the Toxic Twenty. Zurich, Switzerland: Blacksmith Institute's and Green Cross Switzerland; 2011. pp. 34-38

[44] Ramshaw JA, Peng YY, Glattauer V, Werkmeister JA. Collagens as biomaterials. Journal of Materials Science. Materials in Medicine. 2009;**20**:3-8. DOI: 10.1007/s10856-008-3415-4

[45] Muyonga JH, Cole CGB, Duodu KG. Fourier transform infrared (FTIR) spectroscopic study of acid soluble collagen and gelatin from skins and bones of young and adult Nile perch (Lates niloticus). Food Chemistry. 2004;**86**:325-332. DOI: 10.1016/j.foodchem.2003.09.038

[46] Cheng FY, Hsu FW, Chang HS, Lin LC, Sakata R. Effect of different acids on the extraction of pepsin-solubilised collagen melanin from silky fowl feet. Food Chemistry. 2009;**113**:563-567. DOI: 10.1016/j.foodchem.2008.08.043

[47] Kurt A. Investigation of plant derived substances that promote the biosynthesis of collagen; development of cosmetic and dermatological skin care products [PhD thesis]. Ankara: Turkey Middle East Technical University; 2015

[48] Orgel JPRO, Antonio JDS, Antipova O. Molecular and structural mapping of collagen fibril interactions. Connective Tissue Research. 2011;**52**:2-17. DOI: 10.3109/03008207.2010.511353

[49] Noorzai S, Verbeek CJR, Lay MC, Swan J. Collagen extraction from various waste bovine hide sources. Waste and Biomass Valorization. 2019. DOI: 10.1007/s12649-019-00843-2

[50] Shoulders MD, Raines RT. Collagen structure and stability. Annual Review of Biochemistry. 2009;**78**:929-958. DOI: 10.1146/annurev.biochem.77.032207.120833

[51] Zhang Z, Li G, Shi BI. Physicochemical properties of collagen, gelatin and collagen hydrolysate derived from bovine limed split wastes. Journal of the Society of Leather Technologists and Chemists. 2006;**90**:23-28

[52] Ferreira AM, Gentile P, Chiono V, Ciardelli G. Collagen for bone tissue regeneration. Acta Biomaterialia. 2012;**8**:3191-3200. DOI: 10.1016/j.actbio.2012.06.014

[53] Robins SP. Fibrillogenesis and maturation of collagens. In: Seibel MJ, Robins SP, Bilezikian JP, editors. Dynamics of Bone and Cartilage Metabolism. London: Academic Press; 1999. p. 31

[54] Kielty CM, Grant ME. The collagen family: Structure, assembly, and organization in the extracellular matrix. In: Royce PM, Steinmann B, editors. Connective Tissue and Its Heritable Disorders. New York: Wiley-Liss; 2002. p. 159

[55] Brodsky B, Persikov AV. Molecular structure of the collagen triple helix. Advances in Protein Chemistry. 2005;**70**:301-339. DOI: 10.1016/S0065-3233(05)70009-7

[56] Birk DE, Bruckner P. Collagen superstuctures. Topics in Current Chemistry. 2005;**247**:185-205

[57] Hulmes DJS. Building collagen molecules, fibrils, and suprafibrillar structures. Journal of Structural Biology. 2002;**137**:2-10. DOI: 10.1006/jsbi.2002.4450

[58] Ricard-Blum S, Dublet B, Van der Rest M. Unconventional collagens. Types VI,VII,VIII, IX, X, XIV, XVI and XIX. In: Sheterline P, editor. Unconventional collagens. Types VI, VII, VIII, IX, X, XIV, XVI and XIX. Oxford, United Kingdom: Oxford University Press; 2000

[59] Gelse K, Poschl E, Aigner T. Collagens-structure, function and biosynthesis. Advanced Drug Delivery Reviews. 2003;**55**:1531-1546. DOI: 10.1016/j.addr.2003.08.002

[60] Ricard-Blum S, Ruggiero F, van der Rest M. The collagen superfamily. The collagen superfamily. In: Brinckmann J, Notbohm H, Müller PK, editors. Collagen. Topics in Current Chemistry. Vol. 247. Berlin, Heidelberg: Springer; 2005

[61] Matmaroh K, Benjakul S, Prodpran T, Encarnacion AB, Kishimura H. Characteristics of acid soluble collagen and pepsin soluble collagen from scale of spotted golden goatfish (Parupeneus heptacanthus). Food Chemistry. 2011;**129**:1179-1186. DOI: 10.1016/j.foodchem.2011.05.099

[62] Greene DM. Use of poultry collagen coating and antioxidants as flavor protection for cat foods made with rendered poultry fat [MSc thesis]. Virginia, United States: Virginia Polytechnic Institute and State University; 2003

[63] Friess W. Review article. Collagen-biomaterial for drug delivery. European Journal of Pharmaceutics and Biopharmaceutics. 1998;**45**:113-136. DOI: 10.1016/S0939-6411(98)00017-4

[64] Woodley DT, Keene DR, Atha T, Huang Y, Lipman K, Li W, et al. Injection of recombinant human type VII collagen restores collagen function on dystrophic epidermolysis bullosa. Nature Medicine. 2004;**10**:693-695

[65] Ottani V, Martini D, Franchi M, Ruggeri A, Raspanti M. Hierarchical structures in fibrillar collagens. Micron. 2002;**33**:587-596. DOI: 10.1016/s0968-4328(02)00033-1

[66] Knupp C, Squire JM. Molecular packing in network-forming collagens. Advances in Protein Chemistry. 2005;**70**:375-403. DOI: 10.1016/S0065-3233(05)70011-5

[67] Nalbat S, Onem E, Basaran B, Yorgancioglu A, Yilmaz O. Effect of finishing density on the physico-mechanical properties of leather. Journal of the Society of Leather Technologists and Chemists. 2016;**100**:84-89

[68] Engel J, Bächinger HP. Structure, stability and folding of the collagen triple helix. In: Brinckmann J, Notbohm H, Müller PK, editors. Collagen. Topics in Current Chemistry. Vol. 247. Berlin, Heidelberg: Springer; 2005

[69] Fratzl P. Collagen Structure and Mechanics. Boston, MA: Springer; 2010. DOI: 10.1007/978-0-387-73906-9

[70] Prockop DJ, Kivirikko KI, Tuderman L, Guzman NA. The biosynthesis of collagen and its disorders (first of two parts). The New England Journal of Medicine. 1979;**301**:13-23. DOI: 10.1056/NEJM197907053010104

[71] Ikoma T, Kobayashi H, Tanaka J, Walsh D, Mann S. Physical properties of type I collagen extracted from fish scales of Pagrus major and Oreochromis niloticas. International Journal of Biological Macromolecules. 2003;**32**:199-204. DOI: 10.1016/s0141-8130(03)00054-0

[72] Benjakul S, Nalinanon S, Shahidi F. Fish collagen. In: Simpson B, editor. Food Biochemistry and Food Processing. 2nd ed. Oxford: Blackwell Publishing; 2012. pp. 365-387

[73] Silvipriya KS, Krishna Kumar K, Bhat AR, Dinesh Kumar B, Anish J,

Panayappan L. Collagen: Animal sources and biomedical application. Journal of Applied Pharmaceutical Science;**201**(5):123-127. DOI: 10.7324/JAPS.2015.50322

[74] Gomez-Guillen MC, Gimenez B, Lopez-Caballero ME, Montero MP. Functional and bioactive properties of collagen and gelatin from alternative sources: A review. Food Hydrocolloids. 2011;**25**:1813-1827. DOI: 10.1016/j.foodhyd.2011.02.007

[75] Hashim P, Ridzwan M, Bakar J, Hashim D. Collagen in food and beverage industries. International Food Research Journal. 2015;**22**:1-8

[76] Cortial D, Gouttenoire J, Rousseau CF, Ronziere MC, Piccardi N, Msika P, et al. Activation by IL-1 of bovine articular chondrocytes in culture within a 3D collagen-based scaffold. An in vitro model to address the effect of compounds with therapeutic potential in osteoarthritis. Osteoarthritis and Cartilage. 2006;**14**:631-640. DOI: 10.1016/j.joca.2006.01.008

[77] Karim AA, Bhat R. Fish gelatin: Properties, challenges, and prospects as an alternative to mammalian gelatins. Food Hydrocolloids. 2009;**23**:563-576. DOI: 10.1016/j.foodhyd.2008.07.002

[78] Tzaphlidou M, Berillis P. Structural alterations caused by lithium in skin and liver collagen using an image processing method. Journal of Trace and Microprobe Techniques. 2002;**20**:493-504. DOI: 10.1081/TMA-120015611

[79] Liang J, Pei XR, Wang N, Zhang ZF, Wang JB, Li Y. Marine collagen peptides prepared from chum salmon (Oncorhynchusketa) skin extend the life span and inhibit spontaneous tumor incidence in Sprague-Dawley rats. Journal of Medicinal Food. 2010;**13**(4):757-770. DOI: 10.1089/jmf.2009.1279

[80] O'Sullivan A, Shaw NB, Murphy SC, Van de Vis JW, Van Pelt-Heerschap H, Kerry JP. Extraction of collagen from fish skins and its use in the manufacture of biopolymer films. Journal of Aquatic Food Product Technology. 2008;**15**:21-32. DOI: 10.1300/J030v15n03_03

[81] Krishn An ST, Perumal P. Preparation and biomedical characterization of jellyfish (Chrysaora Quinquecirrha) collagen from southeast coast of India. International Journal of Pharmacy and Pharmaceutical Sciences. 2013;**5**:698-701

[82] Chinh NT, Manh VQ, Trung VQ, Lam TD, Huynh MD, Tung NQ, et al. Characterization of collagen derived from tropical freshwater carp fish scale wastes and its amino acid sequence. Natural Product Communications. 2019;**14**:1-12. DOI: 10.1177/1934578X19866288

[83] Mohammadi R, Mohammadifar MA, Mortazavian AM, Rouhi M, Ghasemi JB, Delshadian Z. Extraction optimization of pepsin-soluble collagen from eggshell membrane by response surface methodology (RSM). Food Chemistry. 2016;**190**:186-193. DOI: 10.1016/j.foodchem.2015.05.073

[84] Chaudhary R, Pati A. Poultry feed based on protein hydrolysate derived from chrome-tanned leather solid waste: Creating value from waste. Environmental Science and Pollution Research. 2016;**23**:8120-8124. DOI: 10.1007/s11356-016-6302-4

[85] Schmidt MM, Dornelles RCP, Mello RO, Kubota EH, Mazutti MA, Kempka AP, et al. Collagen extraction process. International Food Research Journal. 2016;**23**:913-922

[86] Skierka E, Sadowska M. The influence of different acids and pepsin on the extractability of collagen from the skin of Baltic cod (Gadus morhua). Food Chemistry. 2007;**105**:1302-1306. DOI: 10.1016/j.foodchem.2007.04.030

[87] Zhang Y, Liu WT, Li GY, Shi B, Miao YQ, Wu XH. Isolation and partial characterization of pepsin-soluble collagen from the skin of grass carp (Ctenopharyngodon idella). Food Chemistry. 2007;**103**:906-912. DOI: 10.1016/j.foodchem.2006.09.053

[88] Liu DS, Liang L, Regenstein JM, Zhou P. Extraction and characterization of pepsin-solubilized collagen from fins, scales, skins, bones and swim bladders of bighead carp (Hypophthalmichthys nobilis). Food Chemistry. 2012;**133**:1441-1448. DOI: 10.1016/j.foodchem.2012.02.032

[89] Dang XG, Yuan HC, Shan ZH. An eco-friendly material based on graft copolymer of gelatin extracted from leather solid waste for potential application in chemical sand-fixation. Journal of Cleaner Production. 2018;**188**:416-424. DOI: 10.1016/j.jclepro.2018.04.007

[90] Anal AK, Noomhorm A, Vongsawasdi P. Protein hydrolysates and bioactive peptides from seafood and crustacean waste: Their extraction, bioactive properties and industrial perspectives. Materials Science. 2013:709-730. DOI: 10.1002/9781118375082.ch36

[91] Liu DS, Wei GM, Li TC, Hu JH, Lu NY, Regenstein JM, et al. Effects of alkaline pretreatments and acid extraction conditions on the acid-soluble collagen from grass carp (Ctenopharyngodon idella) skin. Food Chemistry. 2015;**172**:836-843. DOI: 10.1016/j.foodchem.2014.09.147

[92] Simon A, Vandanjon L, Levesque G, Bourseau P. Concentration and desalination of fish gelatin by ultrafiltration and continuous diafiltration processes. Desalination. 2002;**144**:313-318. DOI: 10.1016/S0011-9164(02)00333-8

[93] Schrieber R, Gareis H. Gelatine Handbook: Theory and Industrial Practice. Eberbach, Germany: Wiley-VCH Verlag GmbH & Co. KGaA; 2007. p. 268

[94] Tkaczewska J, Wielgosz M, Kulawik P, Zając M. The effect of drying temperature on the properties of gelatin from carps (*Cyprinus carpio*) skin. Czech Journal of Food Sciences. 2019;**37**:246-251. DOI: 10.17221/128/2018-CJFS

[95] Veeruraj A, Arumugam M, Ajithkumar T, Balasubramanian T. Isolation and characterization of collagen from the outer skin of squid (Doryteuthis singhalensis). Food Hydrocolloids. 2015;**43**:708-716. DOI: 10.1016/j.foodhyd.2014.07.025

[96] Ali ME, Sultana S, Hamid SBA, Hossain MM, Yehya WA, Kader A, et al. Gelatin controversies in food, pharmaceuticals and personal care products: Authentication methods, current status and future challenges. Critical Reviews in Food Science and Nutrition. 2018;**58**:1495-1511. DOI: 10.1080/10408398.2016.1264361

[97] Khong NMH, Yusoff FM, Jamilah B, Basri M, Maznah I, Chan KW, et al. Nutritional composition and total collagen content of three commercially important edible jellyfish. Food Chemistry. 2016;**196**:953-960. DOI: 10.1016/j.foodchem.2015.09.094

[98] Cho SH, Jahncke ML, Chin KB, Eun JB. The effect of processing conditions on the properties of gelatin from skate (Raja Kenojei) skins. Food Hydrocolloids. 2006;**20**:810-816. DOI: 10.1016/j.foodhyd.2005.08.002

[99] Sow LC, Yang HS. Effects of salt and sugar addition on the physicochemical properties and nanostructure of fish gelatin. Food Hydrocolloids. 2015;**45**:72-82. DOI: 10.1016/j.foodhyd.2014.10.021

[100] Grundy HH, Reece P, Buckley M, Solazzo CM, Dowle AA, Ashford D,

et al. A mass spectrometry method for the determination of the species of origin of Gelatine in foods and pharmaceutical products. Food Chemistry. 2016;**190**:276-284. DOI: 10.1016/j.foodchem.2015.05.054

[101] Mariod AA, Adam HF. Review: Gelatin, source, extraction and industrial applications. Acta Scientiarum Polonorum. Technologia Alimentaria. 2013;**12**:135-147

[102] Lima CA, Campos JF, Filho JLL, Converti A, Carneiro-da-Cunha MG, Porto ALF. Antimicrobial and radical scavenging properties of bovine collagen hydrolysates produced by *Penicillium aurantiogriseum* URM 4622 collagenase. Journal of Food Science and Technology. 2015;**52**:4459-4466. DOI: 10.1007/s13197-014-1463-y

[103] Samad NABA, Sikarwar AS. Collagen: New dimension in cosmetic and healthcare. International Journal of Biochemistry Research & Review. 2016;**14**:1-8. DOI: 10.9734/IJBCRR/2016/27271

[104] Song E, Yeon Kim S, Chun T, Byun HJ, Lee YM. Collagen scaffolds derived from a marine source and their biocompatibility. Biomaterials. 2006;**27**:2951-2561. DOI: 10.1016/j.biomaterials.2006.01.015

[105] Schagen SK. Topical peptide treatments with effective anti-aging results. Cosmetics. 2017;**4**:1-14. DOI: 10.3390/cosmetics4020016

[106] Yorgancioglu A, Bayramoglu EE. Production of cosmetic purpose collagen containing antimicrobial emulsion with certain essential oils. Industrial Crops and Products. 2013;**44**:378-382. DOI: 10.1016/j.indcrop.2012.11.013

[107] Available from: https://www.gelatine.org/applications/beauty.html [Accessed: 01 April 2020]

[108] Xiong XY, Liang J, Xu YQ, Liu J, Liu Y. The wound healing effects of the tilapia collagen peptide mixture TY001 in streptozotocin diabetic mice. Journal of the Science of Food and Agriculture. 2020. DOI: 10.1002/jsfa.10104

[109] Fan J, Zhuang Y, Li B. Effects of collagen and collagen hydrolysate from jellyfish umbrella on histological immunity changes of mice photoaging. Nutrients. 2013;**5**:223-233. DOI: 10.3390/nu5010223

[110] Sibilla S, Godfrey M, Brewer S, Budh-Raja A, Genovese L. An overview of the beneficial effects of hydrolysed collagen as a neutraceutical on skin properties: Scientific background and clinical studies. The Open Nutraceuticals Journal. 2015;**8**:29-42. DOI: 10.2174/1876396001508010029

[111] Lee CH, Singla A, Lee Y. Biomedical application of collagen. International Journal of Pharmaceutics. 2001;**221**:1-22. DOI: 10.1016/S0378-5173(01)00691-3

[112] Yu SM, Li Y, Kim D. Collagen mimetic peptides: Progress towards functional applications. Soft Matter. 2011;**7**:7927-7938. DOI: 10.1039/C1SM05329A

[113] Oliveira SM, Ringshia RA, Legeros RZ, Clark E, Terracio L, Teixeira CC. An improved collagen scaffold for skeletal regeneration. Journal of Biomedical Materials Research. Part A. 2010;**94**:371-379. DOI: 10.1002/jbm.a.32694

[114] Muthukumar T, Sreekumar G, Sastry TP, Chamundeeswari M. Collagen as a potential biomaterial in biomedical applications. Reviews on Advanced Materials Science. 2019;**53**:29-39. DOI: 10.1515/rams-2018-0002

[115] Rethinam S, Vijayan S, Aruni AW, Basaran B, Alagumuthu T,

Ramamoorthy R. Enhanced tissue regeneration using an nano-bioactive scaffold–A novel perspective. Materials Chemistry and Physics. 2020;**240**:1-6. DOI: 10.1016/j.matchemphys.2019.122303

[116] Leitinger B, Hohenester E. Mammalian collagen receptors. Matrix Biology. 2007;**26**:146-155. DOI: 10.1016/j.matbio.2006.10.007

[117] Jeevithan E, Qingbo Z, Bao B, Wu W. Biomedical and pharmaceutical application of fish collagen and gelatin: A review. The Journal of Nutrition. 2013;**2**:218-227

[118] Gorgieva S, Kokol V. Collagen-vs. gelatine-based biomaterials and their biocompatibility: Review and perspectives. In: Biomaterials Application for Nanomedicine. Rijeka: InTech; 2011. pp. 17-52. DOI: 10.5772/24118

[119] Buttafoco L, Kolkman NG, Poot AA, Dijkstra PJ, Vermes I, Feijen J. Electrospinning collagen and elastin for tissue engineering small diameter blood vessels. Journal of Controlled Release: Official Journal of the Controlled Release Society. 2005;**101**:322-324

[120] Friess W, Uludag H, Foskett SM, Biron RM, Sargeant C. Characterization of absorbable collagen sponges as rhBMP-2 carriers. International Journal of Pharmaceutics. 1999;**187**:91-99. DOI: 10.1016/S0378-5173(99)00174-X

[121] Meyer M, Trommer K. Soft collagen-gelatine sponges by convection drying. Brazilian Archives of Biology and Technology. 2015;**58**:109-117. DOI: 10.1590/S1516-8913201400139

[122] Manninen AH. Protein hydrolysates in sports nutrition. Nutrition and Metabolism. 2009;**6**:38. DOI: 10.1186/1743-7075-6-38

[123] Close GL, Sale C, Baar K, Bermon S. Nutrition for the prevention and treatment of injuries in track and field athletes. International Journal of Sport Nutrition and Exercise Metabolism. 2019;**29**:189-197. DOI: 10.1123/ijsnem.2018-0290

[124] Mezenova NY, Verkhoturov VV, Volkov VV, Baydalinova LS, Mezenova OY. Determination of technological parameters of powdery active peptides from fish scales as part of bioproduct for sport nutrition. Izvestiya Vuzov-prikladnaya Khimiya I Biotekhnologiya. 2016;**6**:104-114. DOI: 10.21285/2227-2925-2016-6-2-104-114

[125] Nasri M. Advances in Food and Nutrition Research, Chapter 4, Protein Hydrolysates and Biopeptides: Production, Biological Activities, and Applications in Foods and Health Benefits. Vol. 81. Cambridge, United States: Elsevier; 2017. pp. 109-159

[126] Clark KL, Sebastianelli W, Flechsenhar KR, Aukermann DF, Meza F, Millard RL, et al. 24-week study on the use of collagen hydrolysate as a dietary supplement in athletes with activity-related joint pain. Current Medical Research and Opinion. 2008;**24**:1485-1496. DOI: 10.1185/030079908X291967

[127] Shaw G, Lee-Barthel A, Ross ML, Wang B, Baar K. Vitamin C-enriched gelatin supplementation before intermittent activity augments collagen synthesis. The American Journal of Clinical Nutrition. 2017;**105**:136-143. DOI: 10. 3945/ajcn.116.138594

[128] Jianguo M. Nutrition of Daling cod skin polypeptide liquid and its anti-sports fatigue effect. Archivos Latinoamericanos de Nutrición. 2020;**70**:11-18

[129] Swarnalatha S, Ganesh Kumar A, Tandaiah S, Sekaran G. Efficient and safe disposal of chrome shavings

discharged from leather industry using thermal combustion. Journal of Chemical Technology and Biotechnology. 2009;**84**:751-760. DOI: 10.1002/jctb.2108

[130] Majee S, Halder G, Mandal T. Formulating nitrogen-phosphorous-potassium enriched organic manure from solid waste: A novel approach of waste valorization. Process Safety and Environment Protection. 2019;**132**:16-168. DOI: 10.1016/j.psep.2019.10.013

[131] Taylor MM, Cabeza LF, Brown EM, Manner WN. Chemical modification of protein products isolated from chromium-containing solid tannery waste and resultant influence on physical and functional properties. Journal of the American Leather Chemists Association. 1999;**94**:171-181

[132] Kolomaznik K, Mladek M, Langraaier F, Janáčová D, Taylor MM. Experience in industrial practice of enzymatic dechromation of chrome shavings. Journal of the American Leather Chemists Association. 2000;**95**:55-63

[133] Sirbu C, Cioroianu T, Dumitrascu M. New fertilizers with protein structure with fitostimulator role. Seria Agronomie. 2009;**52**:473-478

[134] Gaidau C, Niculescu M, Stepan E, Taloi D, Filipescu L. Additives and advanced biomaterials obtained from leather industry by-products. Revista de Chimie. 2009;**60**:501-507

[135] Lima DQ, Oliveira LCA, Bastos ARR, Carvalho GS, Marques JGSM, Carvalho JG, et al. Leather industry solid waste as nitrogen source for growth of common bean plants. Applied and Environmental Soil Science. 2010;**2010**:1-7. DOI: 10.1155/2010/703842

[136] De Oliveira DQL, Carvalho KTG, Bastos ARR, de Oliveira LCA, De Sa e Melo Marques JG, De Melo Pereira do Nascimento RS. Use of leather industry residues as nitrogen sources for elephantgrass. Revista Brasileira de Ciência do Solo. 2008;**32**:417-424

[137] Kolomaznik K, Viswanathan M. US/RAS/92/120 Regional Programme for Pollution control in the Tanning Industry in South-East Asia. Enzymatic Digestion of Chrome Shavings. UNIDO Report; 2001

[138] Sudha PN, Latha S, Bharghavi VLN. Towards chrome free chicken-a pilot scale study to remove chromium from leather waste, a source for poultry feed manufacture. Environmental Science: An Indian Journal. 2011;**6**:107-111

[139] Langmaier F, Stibora M, Mladek M, Kolomaznik K. Gel-sol transitions of chrome tanned leather waste hydrolysate. Journal of the Society of Leather Technologists and Chemists. 2001;**85**:100-105

[140] Available from: https://www.gelatine.org/applications/animal-feed.html [Accessed: 01 April 2020]

[141] Phillips WP. Handbook of Hydrocolloids. 2nd ed. Woodhead Publishing Series in Food Science, Technology and Nutrition; 2009. p. 162

[142] Teklay A, Gebeyehu G, Getachew T, Yaynshet T, Sastry TP. Preparation of value added composite boards using finished leather waste and plant fibers—A waste utilization effort in Ethiopia. Clean Technologies and Environmental Policy. 2017;**19**:1285-1296. DOI: 10.1007/s10098-016-1327-4

[143] Vasudevan N, Ravindran AD. Biotechnological process for the treatment of fleshing from tannery industries for methane generation. Current Science. 2007;**93**:1492-1494

[144] Shanmugam P, Horan NJ. Optimising the biogas production from

leather fleshing waste by co-digestion with MSW. Bioresource Technology. 2009;**100**:4117-4120. DOI: 10.1016/j.biortech.2009.03.052

[145] Ammasi R, Sundar VJ, Chellan R, Muralidharan C. Amino acid enriched proteinous wastes: Recovery and reuse in leather making. Waste and Biomass Valorization. 2019. DOI: 10.1007/s12649-019-00912-6

Chapter 8

Characterization of Grafted Acrylamide onto Pine Magnetite Composite for the Removal of Methylene Blue from Wastewater

Kgomotso N.G. Mtshatsheni, Bobby E. Naidoo and Augustine E. Ofomaja

Abstract

Much attention has been focused on chemical modification of natural biomass through grafting. Modification of natural polymers by graft copolymerization has shown to be a promising technique as it functionalizes a biopolymer to its potential, imparting desirable properties onto them. The present study focuses on functional groups such as —CO—NH_2 which were grafted into cellulose from acrylamides. The characterization of the composite was done using Fourier transform infrared spectroscopy (FTIR), X-ray diffraction (XRD), scanning electron microscopy (SEM), Brunauer-Emmett-Teller (BET), thermal gravimetric analysis (TGA), and transmission electron microscopy (TEM). These techniques were used to further demonstrate the formation of the grafted acrylamide composite (GACA). SEM analyses showed existence of strong chemical interactions between pine cone magnetite and acrylamides.

Keywords: characterization, adsorption, acrylamide, methylene blue, grafting

1. Introduction

The rapid development of the textile industry has resulted to a large proportion of industrial wastewater pollution. Methylene blue (MB) dye is a most widely used dye by industries like textile, paper, rubber, plastics, leather, cosmetics, food industries and pharmaceuticals. The textile industry is classified into three main categories, namely; cellulose fibers (cotton, rayon, linen, ramie, hemp and lyocell), protein fibers (wool, angora, mohair, cashmere and silk) and synthetic fibers (polyester, nylon, spandex, acetate, acrylic, ingeo and polypropylene) [1]. The type of dyes and chemicals used in the textile industry are found to differ depending on the fabrics manufactured. Reactive dyes (remazol, procion MX and cibacron F), direct dyes (congo red, direct yellow 50 and direct brown 116), naphthol dyes (fast yellow GC, fast scarlet R and fast blue B) and indigo dyes (indigo white, tyrian purple and indigo carmine) are some of the dyes used to dye cellulose fibers [1]. The textile industry is known to be the main creator of wastewater effluents because it consumes more water for its wet processes. Therefore, globally it is estimated that all wastewater discharge is highly populated. According to the world bank estimation, textile dyeing and finishing treatment given to a fabric generates at least 17–20% of world's industrial wastewater [2, 3].

Dyes often discharged in water effluents contain residues that are highly visible and undesirable even at low concentrations [4]. In addition, they are toxic due to their harmful effects on the human beings. Therefore, it is of vital importance that they are removed from water [5]. Wastewater containing dyes needs to be treated before being discharged into water bodies [6]. Various techniques including chemical oxidation, coagulation-flocculation, membrane processes and biological treatment have shown effectiveness in the removal of methylene blue from waste water [7]. The limitation most of these techniques possess is the incomplete dye removal, poor detection, requirement of expensive equipment and monitoring systems [6]. The performance of adsorption techniques have been applied due to their effectiveness since they remove the entire dye molecule, leaving no fragments in the effluent [8].

Extensive research in recent years has focused on utilizing waste materials from agricultural products (such as pine cones and others) since they are eco-friendly, cost-effective and renewable [9]. Pine cones are naturally occurring agricultural wastes widely found in a plantation in Vanderbijlpark, Gauteng, South Africa. They are of commercial importance and value which is extensively used in different industries [10]. One pine cone consists of 46.5% hemicellulose, 37.4% lignin, 18.8% cellulose and 15.4% extractives [11]. Pine cone powder has been studied extensively in the removal of heavy metal pollutants such as lead, caesium, copper nickel and arsenic from water systems. Activated carbon has been the most employed adsorbent for the removal of dyes due to its outstanding adsorption properties. However, it has limitations by being expensive and it cannot be used in large applications of wastewater treatment. The use of biomass and other microbial cultures in the removal of methylene blue has been extensively studied in recent years. Among others, carbonized organic materials, fly ashes, peat moss, recycled alum sludge, fishery residues and microorganisms such as fungus and algae [12].

The present study reports the development and characterization of grafted pine magnetite composite using grafted acrylamide (GACA) for the removal of methylene blue in wastewater. Grafting is a process of chemically or physically manipulating the surface properties of plant materials such as type and amount of functional groups, surface area and porosity by extraction of plant chemical components in order to improve its adsorptive ability. Grafting of synthetic monomers onto pure biological materials has been successfully performed, e.g., grafting of acrylonitrile onto starch [13] and methyl acrylonitrile onto cotton [14].

2. Materials and methods

2.1 Materials

Pine cones are naturally occurring agricultural wastes found in a plantation in Vanderbijlpark, Gauteng, South Africa. All the chemicals and reagents used throughout this study were of analytical grade reagents and used without any further purification. Acrylicamide, ceric ammonium nitrate (CAN), nitric acid (HNO_3), sodium hydroxide (NaOH), ammonium hydroxide (NH_4OH), ferric sulfate ($FeSO_4$) and methylene blue was supplied by Merck, South Africa. Deionized was used for the preparation of all solutions. The stock solution for methylene blue (1000 mg/L) was prepared by dissolving the required amount of dyes in a 1000 ml of deionized water and the stock solution was further diluted for batch experiments.

2.2 Methods

2.2.1 Synthesis of pine-magnetite composites

A mixture of $FeSO_4·7H_2O$ (2.1 g) and of $Fe(SO_4)_3·XH_2O$ (3.1 g) were dissolved under inert atmosphere in 100 cm^3 of double-distilled water with vigorous stirring. Thereafter, 20 cm^3 of 28% ammonium hydroxide and the appropriate amount of pine powder was added. The reaction was left to run for 45 min at 80°C under constant stirring. The resulting particles, consisting of magnetite attached to the cellulose (hereafter referred to as bio-composite) were washed several times with deionized water and ethanol and dried in a vacuum oven at 60°C overnight. To determine the optimum conditions to achieve the desired products of the biocomposites, we experimented with the following variables: volume of NH_4OH 5, 10, 20, 30, and 40 cm^3; weight of pine powder 1.0, 1.5, 2.0, 2.5, 3.0 and 3.5 g; temperatures 40, 60, 80 and 100°C and reaction times 15, 30, 45 and 60 min.

2.2.2 Synthesis of grafted pine magnetite composite

1 g of pine magnetite composite (PMC), 20 ml of 1.5 M acrylicamide and 135 ml of deionised water were transferred into a three neck round bottom flask at a temperature of 42°C. The reaction was bubbled under nitrogen gas for 30 min to remove the dissolved oxygen under stirring. 10 ml of 0.5 M CAN, dissolved in 0.3 M HNO_3 was slowly added to the reaction to initiate graft co-polymerization and stirring was continued for 2 h. Reaction mass was neutralised by 50% NaOH and precipitated in methanol and thereafter washed with methanol/water (90:10) several times, so that the unreacted PMC and ceric salt were removed. The final residue was dried in a vacuum oven at 40°C.

2.3 Characterization

Qualitative and fundamental identification of the functional chemical groups of grafted pine magnetite composite (GPMC) were carried out with a FTIR (Perkin-Elmer) in the range 450–4000 cm^{-1}. An X'Pert PRO X-ray diffractometer (PAN analytical, PW3040/60 XRD; CuKα anode; λ = 0.154 nm) was used for particle size measurements. The size of the synthesized particles was observed using transmission electron microscope (TEM, FEI TECNAI G^2 SPIRIT) at an accelerating voltage of 150 kV. TGA (Perkin-Elmer (USA) Simultaneous Thermal Analyzer 6000 instrument) was used for determining the weight loss as a function of temperature. Changes in morphology were studied using scanning electron microscopy (SEM), HRSEM Instrument Specs Model: Jeol JSM 7800F field emission scanning electron microscope run operational voltage: 5kVEDS specs Model: Thermo Fischer UltraDry EDS Detector for the graft co-polymerization and incorporation of iron oxide magnetite (Fe_3O_4 PMC).

3. Results and discussions

3.1 FT-IR spectroscopy results

The FT-IR spectrum shown in **Figure 1(a)** represents the pine-Fe_3O_4 magnetite (PMC). The FT-IR spectrum showed some changes in band intensities, indicating the functional groups on the surface that had been modified. A compressed —OH peak at 3350 cm^{-1} with an increase in intensity was observed. This might have been

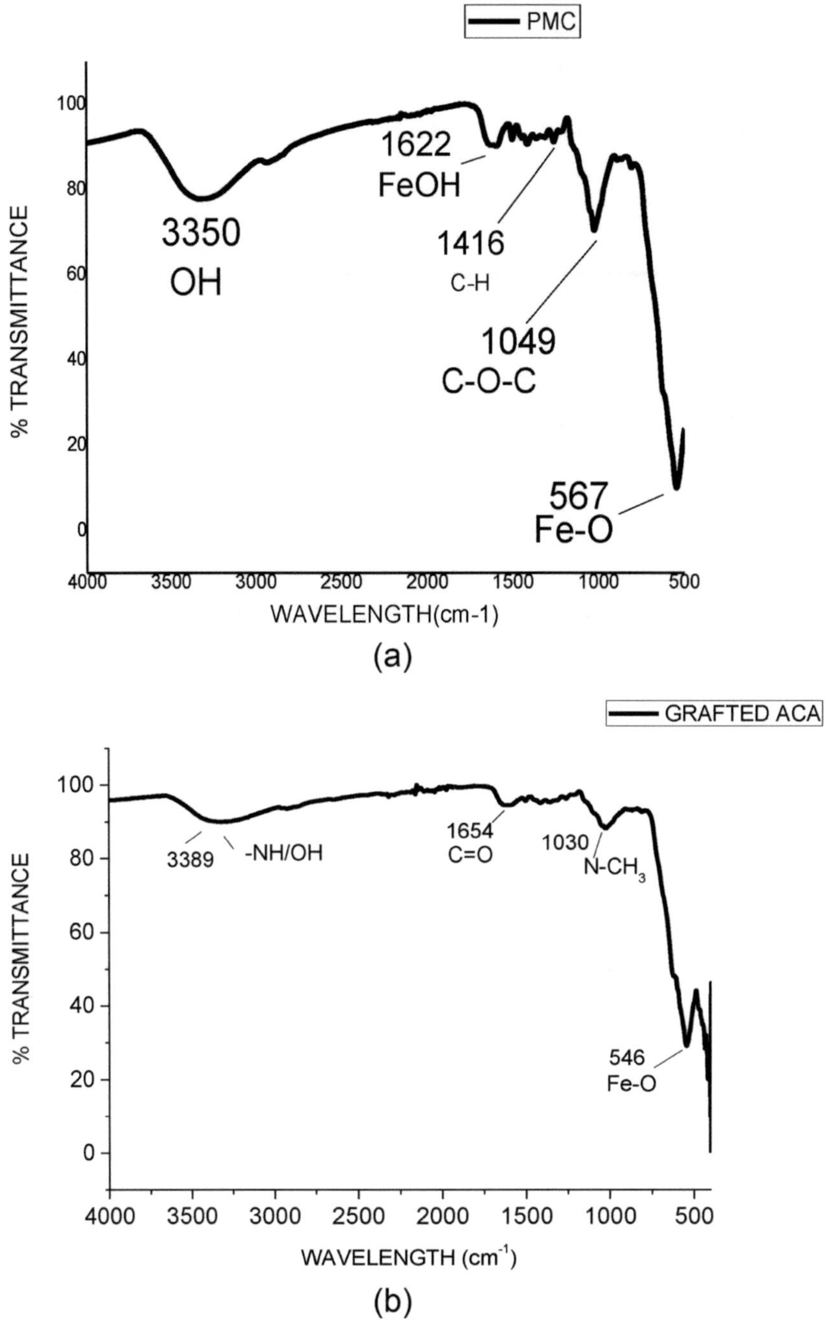

Figure 1.
(a) FT-IR spectrum for PMC and (b) FT-IR spectrum for GACA.

due to the presence of extracted lignin in pine cone. Clearly, the COO— peak was converted to esters at 1622 cm^{-1}, C—H aliphatic peaks were observed at 1416 cm^{-1} which represent the increase in the internal surface of the pine cone and a new peak was found at 567 cm^{-1} which was assigned to the vibration of Fe—O band of Fe_3O_4.

The FT-IR spectrum of GACA (**Figure 1(b)**) shows a slightly broad band observed at 3389 cm^{-1} ascribed to the existence of OH— and —NH groups [15]. The compressed peak at 1654 cm^{-1} corresponds to a carbonyl functional group of acrylamide [16, 17], whereas the one at 1030 cm^{-1} reflects on the vibrations of N—CH_2 groups. The last peak at 546 cm^{-1} reflects the Fe—O functional groups. These functional groups might have participated in the interactions with MB which involved the mechanism of surface complex, hydrogen bonding, and electrostatic attractions.

3.2 XRD analyses

The XRD spectrum of grafted pine magnetite composite with acrylamide is shown in **Figure 2a**. The prominent peaks at 2θ values of 30.5°, 38.7°, 43.9.0°, 59.8° and 63.7° corresponding to (220), (311), (400), (422), (511), respectively, attributes to the cellulose peaks due to the presence of iron oxide magnetite composite and crystal planes of grafted pine magnetic composite respectively [18]. The composite has shown a cubic crystal structure. It is observed that diffraction intensity of the broad peak at 43.9° was weakened indicating that the crystallinity of the PMC decreased after grafting. This phenomenon might be due to the strong interaction of covalent bonds between the PMC and the acrylamide.

3.3 TGA analyses

The TGA and DTG curves shown in **Figure 3a** demonstrate the thermal stability of grafted pine magnetite composite. The incorporation of the Fe_3O_4 magnetite composite showed the changes in the thermal properties of the cellulose. The initial thermal decomposition of GACA occurred at 100–240°C temperature range which corresponds to loss of water molecules and volatile compounds. The second stage thermal decomposition in the temperature range 380–640°C may be due to the breakdown of the polymer matrix and cross-links between different polymeric chains. The last stage of decomposition at a temperature of 700°C corresponds to the lignin degradation [19]. Grafting with acrylamide presented a better thermal stability due to the different types of covalent bonds in the grafting of copolymer backbone [20]. Differential thermal analysis (DTA) showed endothermic peaks associated with degradation of various materials. The degradation behaviour exhibited two stage decomposition effects. Observation at different temperatures (380–620°C) was attributed to the cellulose decomposition at low temperature and grafted acrylamide composite at higher temperatures. This confirmed the stabilizing effect of the incorporation of Fe_3O_4 composite onto acrylamide.

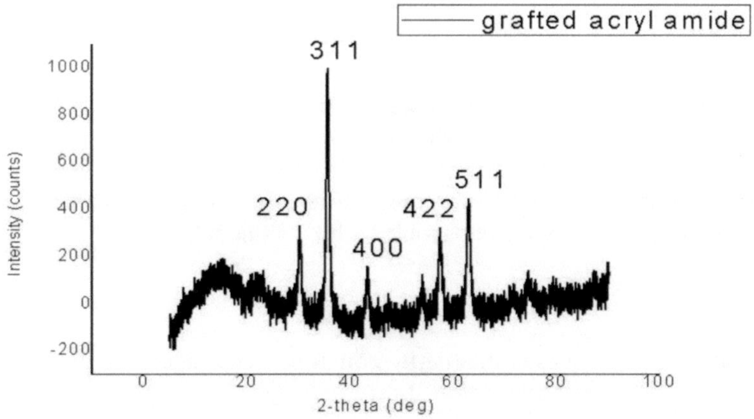

Figure 2.
XRD spectrum for GACA.

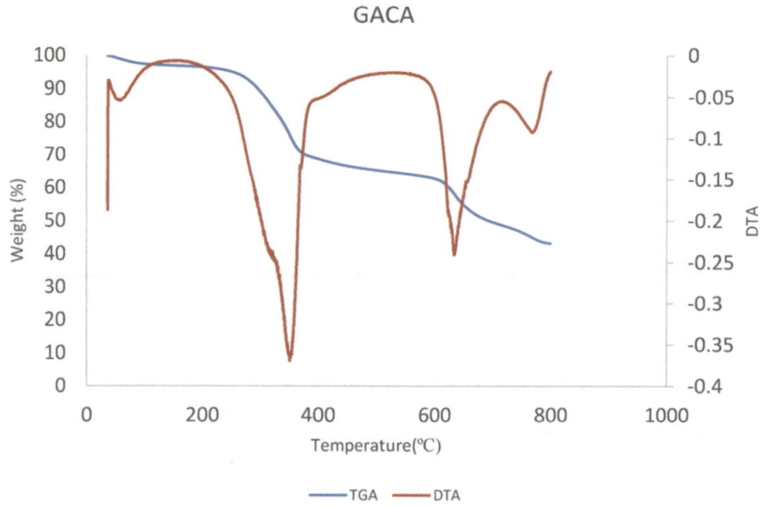

Figure 3.
TGA and DTA curves for GACA.

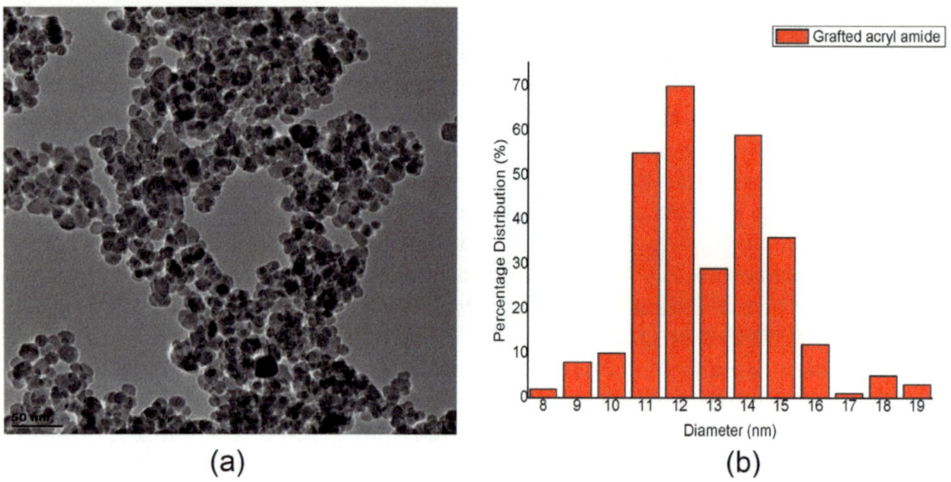

Figure 4.
(a) TEM image and (b) size distribution of GACA.

3.4 TEM studies

The TEM image in **Figure 4(a)** shows the appearance of the typical images of grafted pine magnetic composite with acrylamide. The supporting information showed spherical nano-particles as attributed to the shape and the incorporation of the magnetic nanoparticles in the polymer matrix. **Figure 4(b)** shows the size distribution of the pine magnetite particles with a peak at 12 nm.

3.5 SEM studies

SEM images of grafted pine magnetite composite with acrylamide are shown in **Figure 5(a)** and **(b)**. The observation showed changes in morphology of the GACA

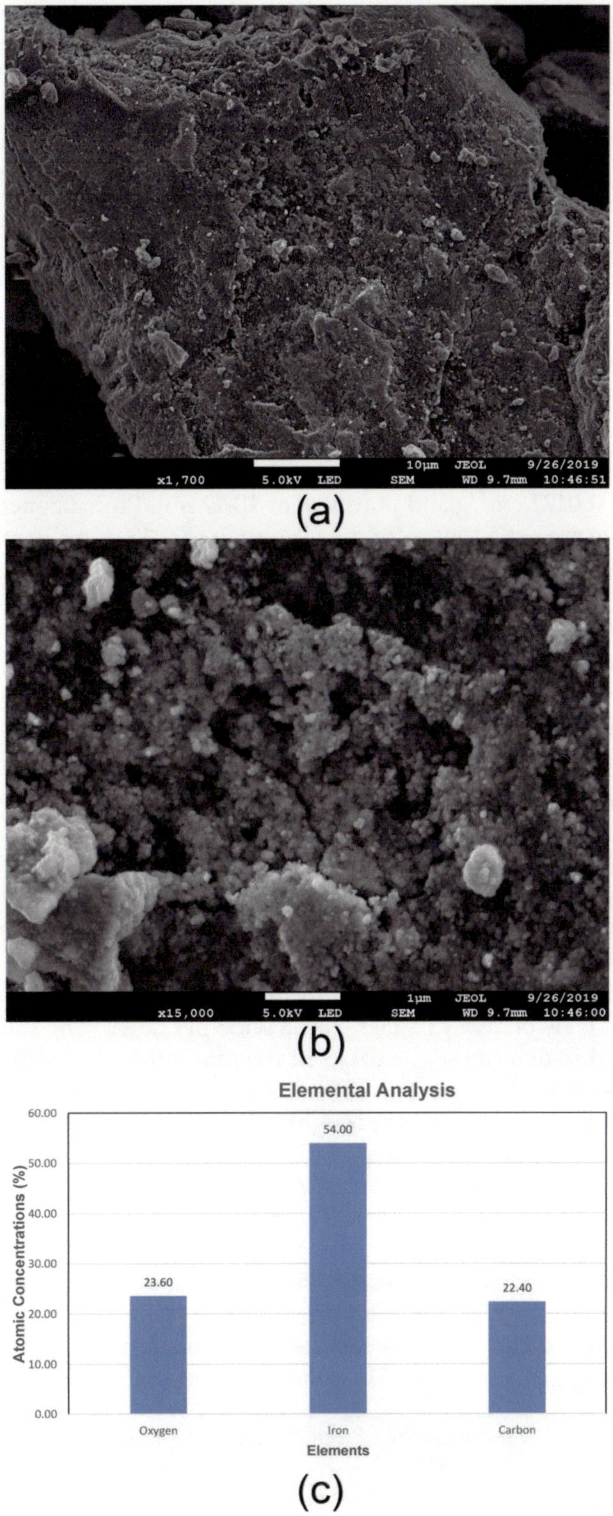

Figure 5.
(a) SEM image of the GACA, (b) SEM image of the GACA and (c) elemental analysis from SEM-EDX.

because of the graft copolymerization process and incorporation iron oxide magnetite. Supporting information showed the granular smooth surface. Roughness of the surface increased after modification, better matrix coherence was achieved after incorporation of the iron oxide magnetite nanoparticles. All the observations confirmed that grafting pine magnetite composite with acrylamide allows better compatibility. The presence of the Fe peak in the EDX of the nanocomposite showed successful incorporation of iron oxide composite in the polymer matrix **Figure 5(c)**.

3.6 BET (surface area) analyses

A surface property of an adsorbent describes the effect of modification on the surface area of the adsorbent. **Table 1** shows comparison of the effect of modification on the surface area of the materials. The pure pine magnetite nanoparticles showed a surface area of 113.60 m^2/g, pore volume of 0.6321 cm^3/g and pore size of 25.86 nm. On the other hand, the NaOH treated pine had a surface area of 2.25 m^2/g, pore volume of 0.0177 cm^3/g and pore size of 10.17 nm. Pine magnetite composite exhibited surface area of 54.80 m^2/g, pore volume of 0.1522 cm^3/g and pore size of 23.10 nm. Grafted acrylamide reflected the surface area of 57.77 m^2/g, pore volume of 0.1591 cm^3/g and pore size of 17.33 nm. The higher surface area was due to the pine cone structure which was found to be important for the improvement of mass diffusion and adsorptive capacity. An increase in surface area, pore volume and pore size confirmed that GACA can adsorb MB more efficiently than the PMC. The distinct pore structure size enables fast transportation of particles.

3.7 Point zero charge (pH_{pzc})

To further investigate the effects of modifications on the suitability of the synthesized materials for adsorption, the isoelectric point or point of zero charge (pH_{pzc}) was determined. The solution pH is an important parameter for dye adsorption because it does not only change the surface charge of the adsorbent but also it affects the molecular structure of the dye. As MB is a cationic dye, it can easily form positively charged species over a wide pH range. The solid addition method was used to determine the pH_{pzc} of the pine cone composite. To a series of 100 cm^3 volumetric flasks, 45 cm^3 of 0.01 mol/dm^3 KNO_3 solution were transferred. The pH_i values of the solutions were roughly adjusted between pH 2 and 12 by the addition of either 0.1 mol/dm^3 HCl or NaOH on a pH meter with constant stirring. The total volume of the solution in each flask was made up to 50 cm^3 by the addition of KNO_3 solution of the same strength. The pH_i of the solutions was accurately noted, and 0.1 g of pine cone composite were added to each volumetric flask, which was then immediately closed. The suspensions were allowed to equilibrate for 48 h on a shaker operating at 200 rpm. The pH_f values of the supernatant were accurately noted and the difference between the initial and final pH values ($\Delta pH = pH_f - pH_i$) were plotted against the pH_i. The solution pH is an important parameter for

Properties	Pure magnetite composite	NaOH treated pine	Pine magnetite composite (PMC)	Grafted acrylamide
Surface area (m^2/g)	113.60	2.25	54.80	57.77
Pore volume (cm^3/g)	0.6321	0.0177	0.1522	0.1591
Ave. pore size (nm)	25.86	10.17	23.10	17.33

Table 1.
BET surface area and pore characteristics for synthesized materials.

dye adsorption because it does not only change the surface charge of an adsorbent, but it also reflects the molecular structure of the dye.

Changes in the point of zero charge values within the sample can be attributed by the difference in types and amounts of surface functional groups present on the surface of the adsorbent. pH_{pzc} is observed when modification on the suitability of the synthesized materials is determined. It is known to be the pH at which the amount of positive charges on a biosorbent surface equals the amount of the negative charge, i.e., the pH at which the biosorbent surface has net electrical neutrality [21, 22]. Methylene blue is a cationic dye and can easily form positively charged species over a wide pH range. The pH_{pzc} of pine magnetite composite was found to be 8.56 and grafted pine magnetite with acrylamide was found to be 6.2. The decrease in the pH_{pzc} is attributed to the modification of the surface area.

3.8 Adsorption studies

3.8.1 Effect of solution pH

The adsorption experiments were carried out using batch equilibration techniques. Various methylene blue (MB) solutions with different pH range, initial concentrations and mass dosage were prepared by diluting 1000 mg/dm^3. Equilibrium experiments, to determine the adsorption capacity of pine magnetite composite were conducted using 250 cm^3 bottles. 0.1 g of PMC and 100 cm^3 of the MB solution were added and shaken for 2 h at 26°C. Thereafter, absorbance was determined using UV-VIS spectrophotometer at the wavelength corresponding to the maximum absorbance (λ_{max} = 665 nm) as determined from the plot. This wavelength was used for measuring the absorbance of residual concentration of MB. pH of the solution was adjusted using 0.1 M HCl and 0.1 M NaOH. **Figure 6** showed the effect of pH on the adsorption of MB. An increase in pH showed an increase in percentage removal. When the pH was 2.0 and 4.0, the removal rate of MB was 99.4 and 99.5%, respectively. This indicated that the lower adsorption of MB at acidic pH was due to the presence of excess H^+ ions. The influence of low pH to MB adsorption was that H^+ ions could occupy the binding sites; this was not favorable for the adsorption of MB. Furthermore, MB possessed positive surface charges and could be repulsed by H^+ ions to prevent MB adsorption onto grafted pine magnetic composite.

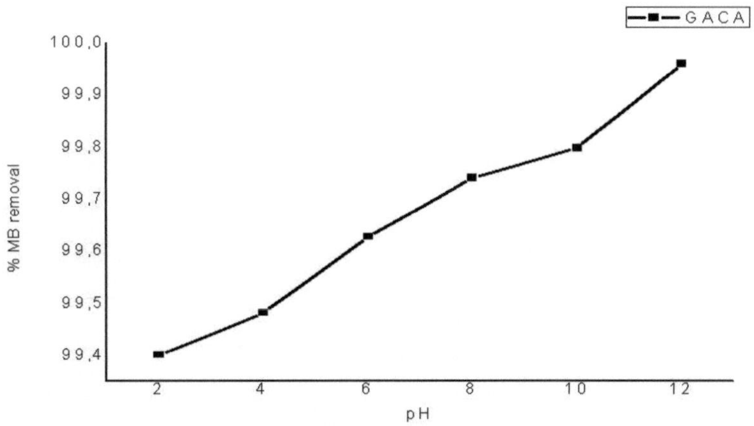

Figure 6.
Effect of pH on the adsorption of MB.

With increasing pH, the number of hydrogen ions i solution was reduced and the competitive effect, repulsive interaction weakened, lead to an increase in the removal rate. The MB removal rate became stable when the pH reached 12, where the higher percentage removal for MB was observed in comparison to other pH values.

3.8.2 Effect of adsorbent dose

Figure 7 shows the effect of adsorbent dose on the percentage removal and amount of dye that was adsorbed. This effect was necessary in order to observe how the novel adsorbent used impacted on the adsorption stoichiometry. It also gave an idea of the propensity of dye molecules to be adsorbed with the smallest amount of adsorbent. When the mass of the adsorbent was 0.5 g, the percentage adsorption removal increased rapidly, which contributed to the increased surface area of the adsorbent which in turn increased the number of binding sites [23]. The adsorption capacity decreased as the amount of adsorbent GACA increased because more active sites were available for the adsorption of dye, which resulted in more interactions between dye and adsorbent thus increasing the MB percentage removal. At mass 0.5 g the highest percentage removal of 99.8% was achieved.

3.8.3 Effect of contact time

The effect of contact time on the grafted pine magnetite composite with acrylamide for the adsorption of methylene blue is shown in **Figure 8**. The adsorption experiment was done at 100 mg/L concentration. The adsorption rate of the grafted composite on the removal of MB is faster from the beginning which might be influenced by the grafted composite with higher specific gravity which makes them better in dispersity and more efficient contact with MB. The adsorption capacity of the grafted composite is higher due to its high surface area.

3.9 Adsorption isotherms

The adsorption isotherm explains the relationship between an adsorbate in the liquid phase and the adsorbate adsorbed on the surface of the adsorbent at

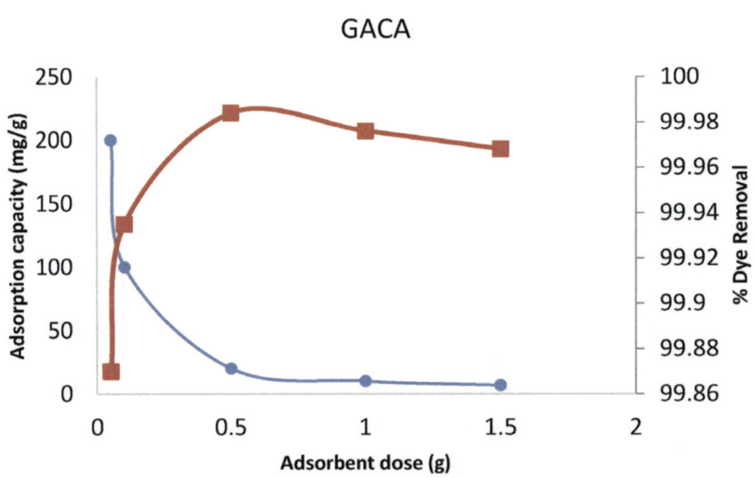

Figure 7.
Effect of adsorbent dose on the adsorption of MB.

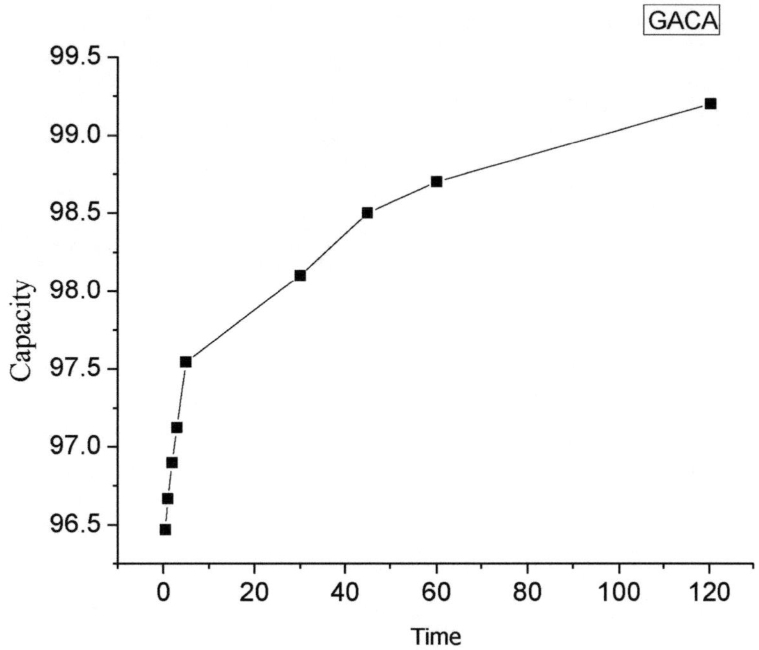

Figure 8.
Effect of contact time at 100 mg/L on the MB adsorption of the GACA.

equilibrium at constant temperature [24, 25]. To successfully obtain the adsorptive behaviour of any substance from the liquid to the solid phase, it is important to have a satisfactory description of the equilibrium state between two phases composing the adsorption system. Langmuir and Freundlich isotherms are the well-known isotherms which have been used to describe the equilibrium of adsorption systems. Typically, the Langmuir model describes the monolayer sorption on a surface containing a limited number of sites and predicting a homogeneous distribution of sorption energies [25]. Freundlich describe the heterogeneity distribution. The results of the MB concentration dependence study were subjected to analyses by using Langmuir and Freundlich isotherm models.

The theoretical Langmuir isotherm is represented by the following equation:

$$C_e/q_e = 1/q_m K_L + C_e/q_m \tag{1}$$

where q_e is the amount of dye adsorbed at the equilibrium time (mg/g), C_e is the equilibrium dye concentration (dm^3/mg), q_m is the maximum adsorption capacity (mg/g) and K_L is the Langmuir adsorption equilibrium constant (dm^3/mg). Freundlich linear expression was represented by:

$$\text{Log}\, q_e = \log K_F + 1/n \log C_e \tag{2}$$

where K_F is the equilibrium adsorption coefficient (dm^3/mg) and $1/n$ is an empirical constant. The parameters of the isotherm models are calculated from the experimental data and the values of correlation coefficient (R^2) are demonstrated in **Table 2**. The results show R^2 values for the Langmuir are higher than those of Freundlich isotherm model. This implies that the equilibrium adsorption data comply with the Langmuir isotherm, suggesting that the adsorption process occurs in a homogeneous surface. Also, it can be stated that the results demonstrate no interaction and transmigration of dyes in the plane of the neighboring surface [26].

Langmuir isotherm model				Freundlich isotherm model		
Temperature (K)	Q_m (mg/m)	K_L (dm^3/mg)	R^2	K_f (mg/g) (dm^3/mg)	N	R^2
299	57.47	0.2107	0.9957	17.7174	2.862	0.9819
304	82.64	0.0588	0.9745	7.66655	1.725	0.9635
309	78.74	0.0968	0.9825	10.7226	1.873	0.9691
314	68.03	0.1786	0.9851	15.01067	2.293	0.9097
319	67.11	0.2569	0.9925	18.02603	2.455	0.8879

Table 2.
Isotherm parameters for methylene blue dye adsorption on GACA.

Higher K_f value for GACA indicates a higher adsorption capacity for methylene blue and a value of $n > 1$ indicates favorable adsorption conditions [27, 28].

3.10 Desorption and regeneration

The main goal of desorption studies is the competitiveness of adsorbents reusability in the multiple adsorption or desorption cycles and their beneficial potential in practical and economical applications. Desorption studies were performed with 0.01 M, 0.05 M and 0.1 M HCl. Typically, 1 g of PMC saturated with 100 mg/L of MB was placed in different desorption solutions and constantly stirred in a water bath at 200 rpm for 2 h. The adsorbent solutions were centrifuged and analysed using UV-VIS spectrophotometer. **Figure 9(a)** demonstrates the effect of eluent concentrations on MB dye desorption efficiency. It was observed that desorption efficiency increased with increase in the eluent concentration even though the shift is small in percentage. The maximum desorption percentage was found at 0.1 M HCl (99.8%) whereby 0.01 M HCl showed the minimum desorption efficiency (98.8%). An increase in HCl concentration resulted in an increase in H^+ ions concentration which led to a subsequent increase in dye desorption efficiency.

Regeneration shows the competitiveness of the adsorbent where it expresses the good reusability and recycling abilities. **Figure 9(b)** demonstrates the possibility of regeneration and reusability of the grafted pine magnetite composite with acrylamide. Adsorption-desorption reaction cycles were repeated 4 times using 0.1 M HCl

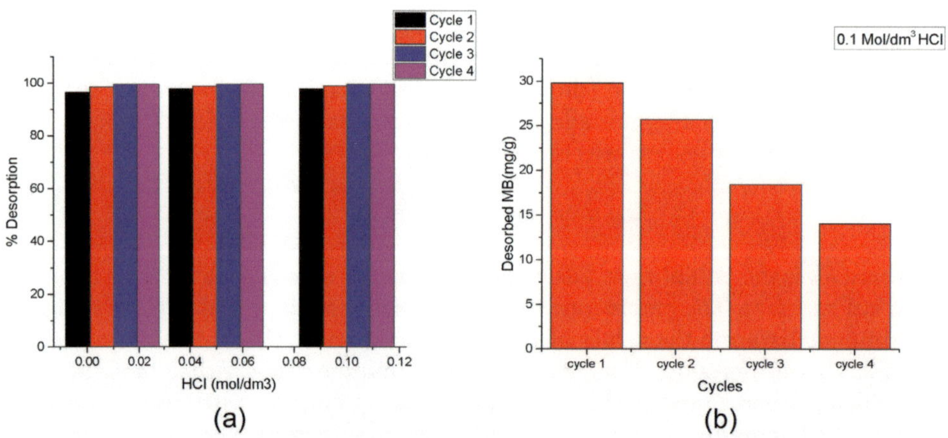

Figure 9.
(a) Desorption characteristics of MB adsorbed and (b) regeneration characteristics of MB adsorbed.

as the desorbing agent. Observation showed that there was a gradual reduction from 29.8% to 14.03% after cycle 4. The results explain that the higher adsorption capacity proves the adsorbent to be a good adsorbent for the removal of MB.

3.11 Conclusions

The study showed that acrylamide was successfully grafted onto pine magnetite composites. FT-IR, BET, SEM, TEM and XRD characterization provided sufficient evidence to demonstrate the incorporation and distribution of the iron oxide nanoparticles within the polymer matrix. GACA nanocomposites were shown to be effective in the adsorption of methylene blue at a pH of 12. The role of adsorbent dose and contact time demonstrated excellent results in the adsorption of methylene blue due to the increased surface area and high rate of the adsorption were achieved. The adsorption data was adequately interpreted by Langmuir and Freundlich isotherm models respectively. It was found that Langmuir isotherm model gave the best equilibrium fit.

Acknowledgements

We are grateful to Prof E.N. Nxumalo and Mr S.A. Zikalala for their contributions on the study. Vaal University of Technology and National Research Foundation of South Africa are thanked for financial support.

Author details

Kgomotso N.G. Mtshatsheni*, Bobby E. Naidoo and Augustine E. Ofomaja
Department of Chemistry, Vaal University of Technology, Vanderbijlpark, South Africa

*Address all correspondence to: kgomotsom@vut.ac.za

References

[1] Ghaly AE, Ananthashankar R, Alhattab M, Ramakrishnan VV. Production, characterization and treatment of textile effluents: A critical review. Journal of Chemical Engineering & Process Technology. 2014;**5**:1

[2] Holkar CR, Jadhav AJ, Pinjari DV, Mahamuni NM, Pandit AB. A critical review on textile wastewater treatments: Possible approaches. Journal of Environmental Management. 2016; **182**:351-366

[3] Yaseen DA, Scholz M. Textile dye wastewater characteristics and constituents of synthetic effluents: A critical review. International Journal of Environmental Science and Technology. 2019;**16**:1193-1226

[4] Amaral FM, Kato MT, Florêncio L, Gavazza S. Color, organic matter and sulfate removal from textile effluents by anaerobic and aerobic processes. Bioresource Technology. 2014;**163**: 364-369

[5] Kausar A, Iqbal M, Javeda A, Aftab K, Nazli ZH, Bhatti HN, et al. Adsorption using clay and modified clay: A review. Journal of Molecular Liquids. 2018;**256**:395-407

[6] Gbekeloluwa O, James H, Dong H. Biosorption of azure dye with sunflower seed Hull: Estimation of equilibrium, thermodynamic and kinetic parameters. International Journal of Engineering Research and Development. 2014;**10**(6): 26-41

[7] Debnath S, Ballav N, Maity A, Pillay K. Competitive adsorption of ternary dye mixture using pine cone powder modified with β-cyclodextrin. Journal of Molecular Liquids. 2017;**225**:679-688

[8] Sara D, Tushar KS. Review on dye removal from its aqueous solution into alternative cost effective and non-conventional adsorbents. Journal of Chemical and Process Engineering. 2014;**1**:104

[9] Alshabanat M, Alsenani G, Almufarij R. Research article removal of crystal violet dye from aqueous solutions onto date palm fibre. Chemistry. 2013;**9**:1-6

[10] Martin-Lara MA, Blazquez G, Calero M, Almendros AI, Ronda A. Binary biosorption of copper and lead onto pine cone shell in batch reactors and in fixed bed columns. International Journal of Mineral Processing. 2016;**148**: 72-82

[11] Ofomaja AE, Naidoo EB, Modise SJ. Biosorption of copper (II) onto potassium hydroxide treated pine cone powder. Journal of Environmental Management. 2010;**91**(8):1674-1685

[12] Lima EC, Royer B, Vaghetti JPC, Simon NM, da Cunha BM, Pavan FA, et al. Application of Brazilian pine-fruit shell as a biosorbent to removal of reactive red 194 textile dye from aqueous solution. Kinetics and equilibrium study. Journal of Hazardous Materials. 2008;**155**(2008):536-550

[13] Gao JP, Tian RC, Yu JG, Duan ML. Graft copolymers of methyl methacrylate onto canna starch using manganic pyrophosphate as an initiator. Journal of Applied Polymer Science. 1994;**53**:1091

[14] Salam MA. Achieving competitive advantage through managing supply chain excellence: The case of Thai garment industry. Journal of Textile and Apparel, Technology and Management. 2005;**4**:1

[15] Reena S, Rajiv ST, Nagpal AK. Effect of cross-linker and initiator concentration on the swelling behaviour and network parameters of superabsorbent hydrogels based on acrylamide and acrylic acid.

International Journal of Plastics Technology. 2009;**13**:22-37

[16] Mohammad S, Behrouz H. Crosslinked graft copolymer of methacrylic acid and gelatin as a novel hydrogel with pH responsiveness properties. Materials. 2011;**4**:543-552

[17] Bhattacharyya R, Ray SK. Enhanced adsorption of synthetic dyes from aqueous solution by a semi interpenetrating network hydrogel based on starch. Journal of Industrial and Engineering Chemistry. 2014;**20**: 3714-3725

[18] Xu Z, Shen C, Tian Y, Shi XZ, Gao HJ. Organic phase synthesis of monodisperse iron oxide nanocrystals using iron chloride as precursor. Nanoscale. 2010;**2**:1027-1032

[19] Ofomaja AE, Ngema SL, Naidoo EB. Grafting of acrylic acid onto biosorbents: Effect of plant components and initiator concentration. Carbohydrate Polymers. 2012;**90**: 201-209

[20] Princi E, Vicni S, Pedemonte E, Mulas A, Franceschi E, Luciano G. Thermal analysis and characterization of cellulose modified with acrylic acid monomers. Thermochimica Acta. 2005; **425**:173-179

[21] Deng Y, Guo Y, Watson H, Au WC, Shakoury-Elizeh M, Basrai MA, et al. Gga2 mediates sequential ubiquitin-independent and ubiquitin-dependent steps in the trafficking of ARN1 from the trans-Golgi network to the vacuole. The Journal of Biological Chemistry. 2009;**284**(35):23830-23841

[22] Ofomaja AE, Naidoo EB, Modise SJ. Removal of copper(II) from aqueous solution by pine and base modified pine cone powder as biosorbent. Journal of Hazardous Materials. 2009;**168**:909-917

[23] Bhumik M, Leswifi TY, Maity A, Srinivasu VV, Onyango MS. Removal of fluoride from aqueous solution by polypyrole/Fe_3O_4 magnetic nanoparticles. Journal of Hazardous Materials. 2011;**186**:150-159

[24] Li K, Wang X. Adsorptive removal of Pb(II) by activated carbon prepared from Spartina alterniflora: Equilibrium, kinetics and thermodynamics. Bioresource Technology. 2009;**100**(11): 2810-2815

[25] Shaibu SE, Adekola FA, Adegoke HI, Ayanda OS. A comparative study of the adsorption of methylene blue onto synthesized nanoscale zero-valent iron–bamboo and manganese-bamboo composites. Materials. 2014;7: 4493-4507

[26] Foo K, Hameed B. Preparation of oil palm (Elaeis) empty fruit bunch activated carbon by microwave assisted KOH activation for the adsorption of methylene blue. Desalination. 2011;**275**: 302-305

[27] Hameed BH, Mahmoud DK, Ahmad AL. Equilibrium modeling and kinetic studies on the adsorption of basic dye by a low-cost adsorbent: Coconut (*Cocos nucifera*) bunch waste. Journal of Hazardous Materials. 2008;**158**:65-72

[28] Hameed BH. Grass waste: A novel sorbent for the removal of basic dye from aqueous solution. Journal of Hazardous Materials. 2009;**166**:233-238

Chapter 9

Wastewater Treatment Using Imprinted Polymeric Adsorbents

Burcu Okutucu

Abstract

In twenty-first century, numbers of synthetic dyes are used in many industries, for example paper, textile, cosmetic, leather for coloring, vs. The dyeing industries wastes is the most found contaminant to be recognized in wastewater. There are various treatment methods including oxidation processes, biological degradation, membrane filtration and coagulation/flocculation have been studied to treat dyeing wastewater. Unfortunately, these methods are high operational costs, complicated operations and possibility of producing more toxic products. Molecularly imprinted polymers (MIPs) are interesting and alternative polymeric adsorbents that can be applied in wastewater treatment for sample preparation and for the quantification of dyes present in wastewater. Molecular imprinting is a process in which functional and crosslinking monomers are co-polymerized in the presence of the target analyte, the imprint molecule. Initially, the functional monomer forms a complex and, after polymerization, their functional groups are held by the highly crosslinking polymeric structure. Upon leaching of the imprint molecule from the polymer matrix, a polymer with binding sites complementary in size and shape to the imprint molecule is created. MIPs can function under extreme conditions of pH, temperature and complex environment. Also, MIPs present wide recognition due to their stability, ease of production and low-cost potential.

Keywords: molecularly imprinted polymer, dye, solid phase extraction, magnetic molecularly imprinted polymers, adsorbent

1. Introduction

With the growth of human; society, science, technology the world is reaching to new high horizons but the cost which will pay in near future is going to be too high. The bad result of this rapid growth is big environmental pollution problem. One of the biggest risks is the water pollution. The major environmental concern worldwide is water pollution associated with release of many different pollutants [1]. Pollutants released into the environment in wastewater effluent, contaminate freshwater resources and are harmful to humans and the environment. These pollutants were released from many different chemical industries such as textile, dyestuffs, metal plating, paper, mining, fertilizer, battery manufacturing metallurgical, pesticides, fossil fuel, tannery, mining, various plastics. Micropollutants, which are mostly found in wastewater, can be pharmaceuticals, industrial chemicals, pesticides and so on. The textile industries micropollutants released in wastewater and sludge can be dyes, phosphorus, sulfamethazine and heavy metals [2]. The release of these contaminants has been increased because of recent rapid industrialization.

IntechOpen

Many industries such as textile, dyestuffs, paper and plastics which are used dyes, use substantial volumes of water in process and also, they generate a considerable amount of colored wastewater [3]. Dye is found a lot amount of pollutant in wastewater. More than 10,000 commercially available dyes are existed and over 7×10^5 of dyes are produced annually throughout the world [4].

Wastewater which is rich by dyestuffs must be treated by many different techniques to eliminate dyes before they are discharged to the ecosystem because of large content of organic pollutants (polycyclic aromatic hydrocarbons, nonphenols, phatales, so on.). The results of many researches showed that about 2% of dyes produced form textile industries are directly discharged in wastewater. Many of these dyes are hazardous and a big threat to aquatic life due to their toxicity and carcinogenicity [3].

Dyes are typically classified in two ways: chemical composition and application class or end use. Based on the application method, dyes can be divided into acid, azoic, basic, direct, disperse, mordant, reactive, sulfur, and vat dyes. According to chemical compositions, dyes can be divided into azo, nitro, nitroso, diarylmethane, triarylmethane, xanthene, anthraquinoid, acridine, cyanine, quinone-imine, pthalocyanine, and thiazole dyes. The azo and anthraquinone dyes are two major classes of reactive dyes, which is released to wastewater as 90% of all organic pollutants. Due to their interaction with hydroxyl ions in the solution, these dyes are always wasted in the dyeing process, remained in the effluents and are hardly eliminated under aerobic conditions. Also, the azo dyes have stability against microbial attack. Reduction of azo bond frequently releases aromatic amines, which resist further degradation under anaerobic condition. The acute toxicity of azo dyes to humans are rather low according to the European Union criteria. The toxicity is generally not because of the dye itself, but because of its degradation products. Azo dyes can be caused contact dermatitis in the local lymph node assays. However azo dyes can be very toxic to several aquatic organisms including algae, fish. Because of presence of azo dyes in water; viability, reproduction rate, filtration feeding and O_2 consumption of the fishes reduced in the freshwater. Azo dyes also inhibit several biological processes, such as they can inhibit algal photosynthesis by reducing the penetration of light, they inhibit chemical oxygen demand (COD) reduction and respiratory activities of microbial populations. Which are ecologically very important. This may result in inhibition of microbial processes like wastewater treatment systems and natural biogeochemical cycles [2, 5, 6].

Over the last years, different physical, chemical and biological techniques have been developed to remove toxic dyes from wastewater and water reservoirs. All kinds of wastewater containing dyes cannot be treated with one technology. The methods for clean up wastewater are adsorption, electrolysis, ozonation, coagulation/flocculation, physical membrane separation, advanced oxidation processes and biological degradation process. Each technology has some merits and demerits. The demerits of many of these techniques are high cost and the formation of hazardous by products. The problems of some techniques can be listed; the quantification of different dyes related to complicated sample preparation procedures, long analysis times, and the use of large quantities of solvents [7]. Coagulation/flocculation usually generates large amounts of toxic sludge difficult to deal with. Physical membrane separation employing nanofiltration (NF) membrane is easy to bring pollution and lead to decreasing removal efficiency Advanced oxidation processes including chemical oxidation, catalytic degradation and electro chemical treatment consume high energy and are still quite costly for practical application [8]. Biological degradation with activated sludge cannot clean up wastewater due to the biodegradable difficulty of many synthetic dyes. The problem of anionic dyes is highly water soluble and difficult to remove by conventional methods. The nonionic dyes (disperse dyes) do not ionize in an aqueous solution and their fused aromatic ring structure makes them highly resistant to degradation. However, a few cationic dyes like methyl blue can be easily removed by adsorption and advanced oxidation processes [4]. With the realization that a single technique cannot give universal solution, the recent trend in dye wastewater treatment is focused around integration of multiple techniques [7, 9].

Adsorption is a low-cost and effective method for expurgate of dye wastewater. The adsorption process is the best choice for the decolorization of dyes and gives the best results for removal of various types of dissolved coloring materials and also removes the entire dye molecule, leaving no fragments in the effluent. The adsorption has a high treatment efficiency and adsorbents can be regenerated for multiple reuses. The initial dye concentration, solution pH, temperature, contact time and adsorbent dosage are usually the main factors that govern the performance. Generally, removal of dyes from water and wastewater is carried out by adsorption using activated carbon. The disadvantage of activated carbon are expensive regeneration process and the decrease of adsorption capacity after regeneration. The other adsorption material can be nanoparticle adsorbents, low cost waste-based adsorbents and polymeric adsorbents. Polymeric adsorbents have the advantages of high flexibility in design, physical stability, porosity, uniform pore size distribution, high surface area, and chemical stability towards acids and bases, feasible regeneration and thermal durability but they are generally expensive materials [4]. The interesting and urgent research needs for a high-efficiency, low cost, attractive and reusable adsorbent for clean up wastewater [10].

2. Molecularly imprinted polymers (MIPs)

Molecularly imprinted polymers (MIPs) are interesting materials that can be applied in environmental research for sample preparation, to clean up and for the quantification of contaminants present in wastewater. The advantages of MIPs are related to their small size, high surface area, and specific selectivity towards target molecules. Molecular imprinting can be prepared with nanoscale surface recognition sites for target molecules, providing high binding capacities and fast mass transfer rates. MIPs have resistance under extreme conditions of pH, temperature, and complex environment. The preparation route of molecular imprinting is in which functional and crosslinking monomers are co-polymerized in the presence of the target analyte (the imprint molecule). Initially, the functional monomer forms a complex (monomer-target analyte) via covalent or noncovalent interactions and, after polymerization, their functional groups are held by the highly crosslinking polymeric structure. Upon leaching of the imprint molecule from the polymer matrix, a polymer with binding sites complementary in size and shape to the imprint molecule is created. This polymer is now capable of selectively rebinding the imprint molecule [6, 9–11]. In covalent approach; the imprinted molecule is covalently coupled to a polymerizable molecule. The binding of template with monomers are relies on reversible covalent bonds. After copolymerization with crosslinker, the imprint molecule is chemically cleaved from the highly crosslinked polymer with harsh conditions. In non-covalent approach which is usually used for preparing MIP is composed of self-assembly between the template and monomer, followed by a crosslinked co-polymerization. The template molecules interact with monomers (during imprinting procedure and the rebinding) via non-covalent interactions, e.g. ionic, hydrophobic and hydrogen bonding [12].

MIPs have been successfully used as an alternative adsorbent in several analytical fields such as separation of enantiomers in liquid chromatography (LC) or capillary electrochromatography (CEC), binding assays, sensors and solid-phase extraction (SPE). SPE is the most advanced application area of the MIP. In recent years, the number of papers which the MIP is used for SPE sorbents (MISPE) increased MISPE is used in the areas of environmental, food and pharmaceutical analysis as selective sorbents for the extraction or for the clean-up of different classes of compounds from various complex matrices [13]. Concerning more specifically the determination of dyes in wastewater, there were too many examples that mentioned above with MISPE. The application of MIP particles in chromatography, can be packed in a column between two frits and be used off-line or they can also pack in a small column to be coupled on-line with LC. The principle of the extraction on a MIP is the same classical SPE sorbents. The desorption of the analytes is achieved by percolating a solvent in order to disrupt the interactions between the analytes and the MIP. Nevertheless, there are more and more applications of MIPs directly to real wastewater samples without a preliminary treatment [14]. The importance of the washing step and the difficulty to optimize this step was the most important problem of MIP usage.

The successful preparation of MIPs depends on the choice of monomers, the crosslinkers, and the appropriate polymerization conditions. The structure and the functionalities of the template molecule-monomer interaction define the subsequent properties of the binding sites. The criteria to consider when selecting a candidate template molecule are its cost, its availability and its chemical functionalities defining which is the ability to strongly interact with monomers. In non-covalent imprinting, the interactions involved are weak; and the excess amount of template should be used. The main factor which is important for MIP process is choosing the best monomer. The role of the monomer is to provide functional groups which can form a complex with the template by covalent or non-covalent interactions. The strength of the interactions between template and monomer affects the affinity of MIPs and determines the accuracy and selectivity of recognition sites [13]. Many techniques are used to select best suitable monomer. The rational design (computer simulation) and analytical techniques includes nuclear magnetic resonance, UV-vis, Fourier-transform infrared spectroscopy and isothermal titration calorimetry have been studied [15].

The common monomers, which are used for molecular imprinting, are methacrylic acid (MAA), acrylic acid (AA), 2- or 4-vinylpyridine (2- or 4-VP), acrylamide, trifluoromethacrylic acid, 2-hydroxyethyl methacrylate (HEMA), methyl methacrylate, acrylic acid, 2-acrylamido-2-methyl-1-propanesulfonic acid, 4-vinylbenzene boronic acid, 1-vinyl imidazole, allylamine, itaconic acid, urocanic ethyl ester, methacrylamide, acrylonitrile and styrene. MAA named as a "universal" functional monomer due to its unique characteristics, being capable to act as a hydrogen-bond donor and acceptor and showing good suitability for ionic interactions. The monomer is chosen according to functional groups of template molecule. The attempts for finding new functional monomers for synthesis of MIPs continued. Because of their special structures, β-cyclodextrins (β-CDs) can be interesting monomers. The β-CDs are composed of cyclic oligosaccharides with a hydrophilic exterior and a hydrophobic cavity. The molar ratio between template and monomer are important because this ratio affect the affinity and imprinting efficiency of MIPs. The lower molar ratios mean less binding sites in polymers. The over-high ratios mean higher non-specific binding capacity and decrease the binding selectivity. For best imprinting efficiency, the molar ratio of templates to monomers have to be optimized [14, 15].

The role of the cross-linker in imprinted polymer is to organize the functional groups of functional monomers around imprinted molecules, and to form highly rigid polymer The common cross-linkers which is used ethylene glycol dimethacrylate (EGDMA), trimethylolpropane trimethacrylate (TRIM), N,N-methylene bisacrylamide (MBAA) and divinylbenzene (DVB), pentaerythritol triacrylate, 1,4-diacryloyl piperazine [15]. Types and amounts of cross-linkers effects the selectivity and binding capacity of MIPs. If the amount of cross-linkers is too low, MIPs cannot maintain stable cavity configuration. The over-high amounts of cross-linkers will reduce the number of recognition sites.

The solvent is one of the other most important factors determining effective molecular recognition. Because the accuracy of the assembly between the template and the monomer is related to the physical and chemical characteristics of the solvent. The solvent, which is used while preparing MIPs, named as porogen. The porogen acts as not only a solvent but also cavity maker. Aprotic/low polar organic solvents (toluene, acetonitrile, chloroform) are often used in non-covalent polymerization processes in order to obtain good imprinting efficiency. Nevertheless, the MIPs prepared in organic solvent cannot be compatible with aqueous media usage. A few studies reported the use of polar/protic media (methanol, ethanol, water) for the synthesis of MIPs. The development of water-compatible MIPs is an interesting area are going to study [16].

Actually, MIPs are synthesized by free radical polymerization, generated by the thermal or photo homolysis of a chemical bond on an initiator. The most widely used initiators for MIPs synthesis are benzoyl peroxide (BPO), 2,2-dimethoxy-2-phenylacetophenone (DMPA) and 2-azobis(2-methylpropionitrile) (AIBN) and 2,20-azo-bis(2,4-dimethyl) valeronitrile (ABDV) [17].

3. Synthesis methods of MIPs

The synthesis method of MIPs can be a different type. Bulk, suspension, emulsion, precipitation, multi-step swelling, and surface imprinting were the methods which were summarized. Bulk polymerization is the common method to prepare MIPs. The bulk polymerization is rapid, simple preparation, needs no sophisticated or expensive instrumentation. But after polymerization of the bulk polymer, it has to be crushed, ground and sieved to an appropriate size. After grinding, the polymer was in irregular particles in shape and size, and also some high affinity binding sites are destroyed. Bulk polymers have a heterogeneous binding site distribution which is limited the use of MIPs in chromatography. The other techniques of preparing imprinting polymers are suspension polymerization, emulsion polymerization, seed polymerization, precipitation polymerization and surface imprinting. The more homogeneous binding site distribution can be taken by these techniques. The MIPs which is prepared by conventional suspension polymerization, where water is used as a continuous phase to form a droplet of pre-polymerization mixtures in the presence of a stabilizer or surfactant. However, the MIPs prepared by suspension polymerization is polydisperse in size (a few to a few hundred micrometers) and displayed poor recognition. The liquid perfluorocarbon or mineral oil (liquid paraffin) can be used while preparing MIPs by suspension polymerization technique to prevent poor recognition. But, liquid perfluorocarbons immiscible with almost all organic solvents (chloroform, dichloromethane, toluene), monomers and cross-linkers. The emulsion polymerization is as an effective method to produce monodispersed polymeric particles. The disadvantage of emulsion polymerization was the presence of remnants of surfactant in polymerization media. Precipitation polymerization have some advantages in synthesizing spherical particles such as free of surfactant, in one single preparative step and with excellent control over the particle size. The seed polymerization, a typical multi-step swelling and polymerization, produced monodispersed

MIPs. In this technique, the use of water was weakened recognition. In addition, the multistep procedure is very time-consuming. Surface imprinting is another technique nowadays using MIPs mostly synthesized. The biggest problem of MIP was bleaching of template molecule due to the high cross-linking nature of MIPs, which will result in incomplete template removal, and slow mass transfer. Fortunately, this problem can be resolved by surface imprinting, in which the imprinted templates are situated at the surface of the material's surface. Compared to traditional MIPs, surface imprinted polymers possess not only higher binding capacity but also faster mass transfer and binding kinetics. Many particles have been used for the surface imprinting process, such as activated silica, Fe_3O_4 (magnetic) nanoparticles, chitosan, activated polystyrene beads, quantum dots (QDs) and alumina membranes. The magnetic separation is an effective technique for separation of complicated samples. The magnetic nanoparticles have an advantage of its fast recovery, high efficiency, low cost, and direct purification from a mixture without any pretreatment. In recent years, magnetic MIPs have become a hotspot based on the significant advantages of magnetic separation over conventional methods. Generally, preparation of MIPs-coated magnetic nanoparticles (MNPs) involves three steps: (1) preparation of Fe_3O_4 MNPs; (2) surface modification of Fe_3O_4 MNP with TEOS, oleic acid, ethylene glycol or poly (vinyl alcohol) and (3) synthesis of surface imprinted MNPs using a sol–gel process or free radical polymerization [16–18].

4. Examples of MIPs applied for wastewater treatment

Till 1980s, according to MIP database number of MIP paper was published. Eighty percent of these papers were the usage of MIPs as SPE adsorbent. By the usage of nanoparticles for MIPs, the new papers were about the surface imprinting and also to use them in SPE. The some examples were summarized above. The most attractive and useful examples were chosen.

First example was the paper of Deng et al. [8]. Deng et al. was studied Ti(IV) functionalized chitosan molecularly imprinted polymer (Ti-CSMIP). While preparing chitosan imprinted polymer he used Ti^{4+} as Lewis acidic for producing metal hydroxyl group and protonated surface of MIP. He used Ti(IV) functionalized chitosan molecularly imprinted polymer to recognize reactive brilliant red (X-3B) in aqueous solution. The dye behaved as a Lewis base. The MIP was characterized by FTIR, SEM, XRD, BET, elemental and zeta potential analysis. Batch adsorption experiments (sorption isotherm, kînetics) and reusability were performed to evaluate adsorption condition. Regeneration experiments indicated that Ti-CSMIP was an effective sorbent for the selective removal of azo anionic dye in aqueous solutions. The difference of the study was; the Ti-CSMIP used as adsorption and also recognition. By using Ti^{4+} as a Lewis acid, the recognition problem of dye, which is limited due to its macromolecular structure, was overcoming. The Ti(IV) functionalized chitosan MIP was recognized dye macromolecules as template, because X-3B was behaved as a Lewis base. In this study, the first step is preparing Tİ-CSMIP. First of all, chitosan was dissolved and mixed with $Ti(SO_4)_2$ Then $Ti(SO_4)_2$/CS mixture was cross-linked with KH-560. After gelation, the product was treated by microwave for 20 min and the obtained product was washed with distilled water and dried. The product was added into X-3B solution after stirring the product was ready. The X-3B was removed using NaOH and HCl, respectively. Non-imprinted polymer (Ti-CSNIP) was prepared with the same procedure in the absence of the template molecule. Experiment of pH effect indicated X-3B sorption on Ti-CSMIP was dependent on solution. The Temkin and Sips models was used as equilibrium model recycling experiments demonstrated that Ti-CSMIP had a greater potential even after several cycles of regeneration.

The second example was magnetic MIPs which was studied by Luo et al. [19]. He was studied magnetic and hydrophilic molecularly imprinted polymers (mag-MIPs) which were prepared by an inverse emulsion–suspension polymerization technique to remove water-soluble acid dyes from contaminated water. The attractive point of study was using 1-(1-methyl acrylate)-3-methylimidazolium bromide (1-MA-3-MI-Br) as a new functional monomer. The thermal stability, chemical structure and magnetic property of the 1-MA-3-MI-Br-mag-MIPs were characterized by the thermal-gravimetric analyzer (TGA), Fourier transform infrared spectrometer (FT-IR) and vibrating sample magnetometer (VSM), respectively. The first step was preparing novel magnetic nanoparticles and modifying surface by PEG. The final brown suspension was placed for 24 h quiescence, and the supernatant was discarded. The second step was preparing inverse emulsion. According to the method; 2 mL toluene with 0.1 g AIBN was mixed in 50 mL beaker. When AIBN was dissolved, 10 mL TRIM and a drop of Span 80 were added in the mixture and stirred to be uniform. Then 10 mL Fe_3O_4 magnetic fluid was added in the mixture. The mixture was stirred for 5 min, and then the mixture was submerged in the ultrasonic bath for 5-min. Finally, the inverse emulsion can be obtained. The third step was preparing mag-MIPs and magnetic non-imprinted polymers (mag-NIPs) were prepared via inverse emulsion–suspension polymerization. The procedure was summarized, 1 mmol tartrazine and 4 mmol 1-MA-3MI-Br were dissolved and prepolymerized for 30 min and mixed with inverse emulsion. Thirdly, the reaction mixture was heated and purged by N_2 to remove the oxygen. The reaction temperature was at 70°C for 12 h. After the reaction completed, the resulting mag-MIPs with uneluted molecule were filtered by 120 mesh sieve and washed by methanol and 60°C water. Finally, the products were washed by methanol:ammonia solution (9:1, v/v) for 24 h. The mag-NIPs were prepared by the same manner in the absence of template molecules. The several advantages of the water-compatible 1-MA-3-MI-Br-mag-MIPs to remove and recycle the water-soluble acid dyes in water media could be existed. Firstly, the removal efficiency towards water-soluble acid dyes is very high with all above 95% in wastewater. Secondly, the 1-MA-3-MI-Br-mag-MIPs can be reused at least five times without obvious decrease in the removal efficiency. Thirdly, due to the encapsulated Fe_3O_4, the 1-MA-3-MI-Br-mag-MIPs can be easily separated by external magnetic field.

The third example was the combination of MIP with biosensor [20]. Khan et al. was studied a biomimetic sensor which was prepared on carbon paste with magnetic molecularly imprinted polymer (mag-MIP) for sensitive and selective detection of methyl green dye. The mag-MIP was synthesized using a functional monomer that was selected before by computational simulation. The findings showed that imprinted biosensor can be suitable for real samples. The mag-MIP was synthesized by the process of polymerization in the existence of methyl green and the functional monomer acrylamide were dissolved in ethanol The reaction mixture was agitated in a water bath at 25°C for 12 h, followed by addition of Fe_3O_4@SiO_2—C=C and shaking for a further 3 h. In the next step, EGDMA and AIBN were put to the system to the mixture that was further sonicated in a water bath for 5 min. The reaction mixture was thermal polymerized at 60°C for 24 h. The analyte molecule was then removed by Soxhlet extraction apparatus via methanol:acetic acid (9:1, v/v) washing solution. The product obtained (mag-MIP) was dried out at 60°C in vacuum. The magnetic non-molecularly imprinted polymer (mag-MIP) was prepared under the same conditions, but without the presence of the analyte.

This polymer showed higher binding capacity for methyl green, compared to the corresponding non-imprinted polymer. Electrochemical sensors using the mag-MIP produced by this method have some advantages that it was offering mechanical stability in solution and providing satisfactory performance of sensor in terms of sensitivity and selectivity.

The fourth example was different example from the others [7]. In this study, Yu et al. is studied selective, sensitive and reliable magnetic molecularly imprinted material to enrich and separate aromatic amines from azo dyes. The 4,4′-methylenebis(2-chloroaniline) (MOCA) was used as template molecule. The synthesis of magnetic molecularly imprinted polymer was performed on the surfaces of the magnetic silica gel spheres via thermal polymerization as mentioned paper, detailly. The steps can be summarized, $Fe_3O_4@SiO_2$ was prepared and washed sequentially with both ethanol and toluene. The second step was the surface of $Fe_3O_4@SiO_2$ was grafted with vinyl by using VETS and triethylamine. The final product of $Fe_3O_4@SiO_2@\text{-}CH{=}CH_2$ was separated and then dispersed in toluene. The synthesis of magnetic molecularly imprinted nanoparticles was done with EGDMA and 2-VP were used as the cross-linker and functional monomer, respectively. A template molecule of MOCA and 2-VP were dispersed in of toluene. Then, particles $Fe_3O_4@SiO_2@\text{-}CH{=}CH_2$, EDGMA and AIBN mixed with the MOCA+2-VP solution and purged with nitrogen. The thermal polymerization was performed with changing temperature. The non-imprinted particles were synthesized with the same procedures in the absence of MOCA. A solution of methanol and acetic acid (9:1, v/v) was used in a Soxhlet to remove the template molecule. The physical properties of mag-MIPs and NIPs were characterized, and the adsorption isotherms were studied. The adsorption process was described by a pseudo-second order model and the equilibrium data fitted well to a Freundlich equation. The other advantages of these magnetic molecularly imprinted polymers were dye removing from wastewaters effectively, and also discriminating carcinogenic aromatic amines which were the main structure of textile dyes.

The fifth example was one of real sample application [21]. Foguel et al. is studied MIPs for the Acid Green 16 (AG16) textile dye and the used this MIP for rebinding, selectivity and application of in wastewater samples. MIP synthesis was performed using AG16 dye (template), 1-vinylimidazole (functional monomer), ethylene-glycol-dimethacrylate (cross-linker), 2,2′-azobis(2-methylpropionitrile) (initiator) and methanol (solvent) by bulk polymer synthesis. The imprinted polymer presented excellent rebinding of 83%, an imprinted factor of 6.91 and great selectivity in comparison with other textile dyes. Additionally, the MIP showed high efficiency in the extraction of this dye in wastewater and have a better performance when compared to commercial SPE cartridges. In this study the selectivity of the MIP for AG16 was evaluated using four dyes, commonly used in the textile industry, with different structures and chromophore groups: Direct Yellow 50 (DY50), Acid Red 1 (AR1), Basic Red 9 (BR9) and Methyl Green (MG). MIP for AG16 is quite selective compared to the dyes DY50, AR1, MG and BR9, since approximately 86% of AG16 was bound to the MIP, while the binding percentage for these other four dyes was between 4% and 11%. The MIP proposed in this work showed great efficiency in the determination of the AG16 dye, since the synthesized MIP presented good rebinding of the analyte to the selective cavities of the MIP, high selectivity compared to other textile dyes and efficiency in the extraction of the compound of interest, when applied in a sample of textile wastewater.

The sixth example was about removal and also adsorption of dye on MIPs [22]. Okutucu et al. is studied a molecularly imprinted polymer (MIP) which textile dye (Direct Red 23) was used as template for decolorization of textile wastewater and also used for leaching of this dye from the wastewater by adsorption of onto polymer. Acrylamide was used as a monomer and dimethyl sulfoxide was used as a porogen. The ethylene-glycol-dimethacrylate (cross-link), 2,2′-azobis(2-methylpropionitrile) (initiator) was used while preparing bulk polymer by thermal synthesis. The template was removed by methanol:conc. HCl washings. The selective recognition ability of the MIP was studied by an equilibrium-adsorption batch method. The effective adsorption properties of the polymer were tested in synthetic dye wastewater. The high adsorption rate and the amount of imprinted dye that was removed from the polymer was nearly 65%. Eighty percent of the dye was adsorbed by imprinted polymer in synthetic wastewater. The goal of this study is to prepare dye-imprinted polymer to research the molecular recognition characteristics, so that to get valuable results for contamination of dyes in wastewater. It is also important and valuable to recognize dye molecules and their removal from wastewater for the protection of the environment. The dye-imprinted polymer can also be used as a solid-phase adsorbent for Direct Red 23 dye to detect whether it was present in wastewater. Adsorption of dye molecules onto a sorbent can be an effective, low-cost method of decolorization of textile wastewater. Most of the techniques used for this aim were the high cost of production and the regeneration also makes them uneconomical. Molecular imprinting polymers are a new kind of materials which can be economical and effective adsorbents. The removing effect of MIP was seen at **Figure 1**.

The seventh example was the example of using β-cyclodextrin as a monomer for MIP synthesizing [23]. In this study, Hu et al. were synthesized a magnetic β-cyclodextrin polymer (MNP-CM-CDP) which could be used in aqueous media. Kinetic isotherms and a dye adsorption method provided Langmuir. By using MNP-CM-CDP model pollutants (BPA, MB, BO_2, RhB, Cr(III), Pb(II), Zn(II), and Cu(II)) were rapidly and efficiently removed from the aqueous solution. Because of magnetic character, the polymer could be easily separated from the solution under an external magnetic field. The synthesis of the magnetic β-cyclodextrin polymer was done at three steps, with TFTPN used as rigid crosslinker, EPI used as flexible crosslinker, chloroacetic used as carboxymethyl agent, Fe_3O_4 used as magnetic matrix and deionized water used as solvent. First step was, β-TFTPN was dissolved in EPI and then added dropwise to cyclodextrin in NaOH solution. After magnetic stirring for 3 h at 100°C, the solution was filtered. The precipitate was washed with water and THF, and then dried. The second step was synthesis of the CME-CDP. The T-E-CDP was dispersed into NaOH and chloroacetic acid was added. After magnetic stirring, the solution was cooled and neutralized with hydrochloric acid. The CME-CDP was obtained. Third step was synthesis of the MNP-CM-CDP. The magnetic β-cyclodextrin polymer (MNP-CM-CDP) was synthesized by one-step coprecipitation. Fe solutions and the CME-CDP were mixed,

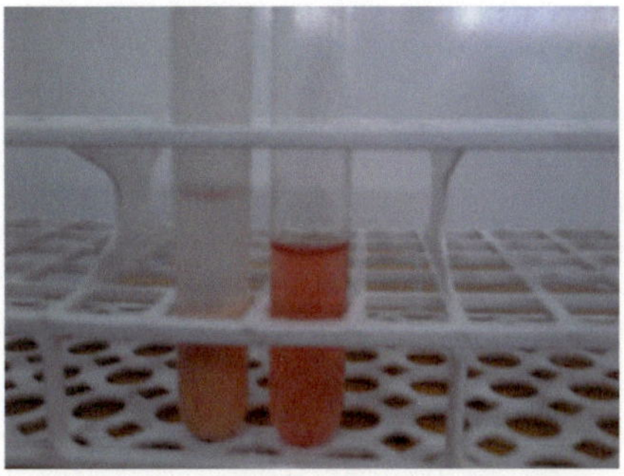

Figure 1.
The removal of Direct Red 23 with MIP [12].

then $NH_3 \cdot H_2O$ (25%, 5 mL) was added dropwise under nitrogen. The precipitate was washed with deionized water five times and then dried. The adsorption of dyes and metal ions is mainly dependent on the pH and this condition was also studied and optimized. The polymerization was performed by green synthesis route, due to use β-cyclodextrin good adsorption performance, easy regeneration, and simple operation. The results indicated that the MNP-CM-CDP is a promising adsorbent in commerce for purifying dyeing wastewater and can be produced on a large scale in industry.

5. Conclusion

Saving water to save the planet and to make the future of human was the most important aim. There are many materials existed that can be used. These are activated carbon, clay minerals, zeolites and biomaterials. The main purpose of all these materials was adsorption of wastes in water. Such as zeolites are crystalline substances with pores which was permitted only the passage of certain size. Zeolites can occur naturally or synthesized in laboratory. They are mainly constituted by Al, O and some metals. They are mostly negatively charged, and their surface can be modified. So, the surface can be treated with multiple classes of contaminants. There are many studies about dye adsorption with different zeolites in wastewater [24, 25]. The zeolites only can remove dyes by adsorption on their surface, but MIPs can ability of recognition, quantification and adsorption. All of these excellent characteristics make MIPs good alternative adsorbents for wastewater treatment. The ease of preparation, resistance to hard conditions (pH, temperature, storage stability, vs.), easily to study with real complex examples (directly wastewater) make MIPs more suitable for different samples. In this chapter, the characteristic components of MIP, synthesizing methods and some examples of treatment of wastewater by MIPs were summarized.

Author details

Burcu Okutucu
Department of Biochemistry, Faculty of Science, Ege University, Bornova, Izmir, Turkey

*Address all correspondence to: burcu.okutucu@ege.edu.tr

References

[1] Gupta VK, Suhas. Application of low-cost adsorbents for dye removal—A review. Journal of Environmental Management. 2009;**90**:2313-2342. DOI: 10.1016/j.jenvman.2008.11.017

[2] Berradi M, Hsissou R, Khudhair M, Assouag M, Cherkaoui O, Bachiri AE, et al. Textile finishing dyes and their impact on aquatic environs. Heliyon. 2019;**5**:e02711. DOI: 10.1016/j.heliyon2019.e02711

[3] Alaba PA, Oladoja NA, Sani YM, Ayodele OB, Mohammed IY, Olupinla SF, et al. Insight into wastewater decontamination using polymeric adsorbents. Journal of Environmental Chemical Engineering. 2018;**6**:1651-1672. DOI: 10.1016/j.jece.2018.02.019

[4] Ahmad A, Mohd-Setapar SH, Chuong CS, Khatoon A, Wani WA, Kumard R, et al. Recent advances in new generation dye removal technologies: Novel search for approaches to reprocess wastewater. RSC Advances. 2015;**5**:30801-30859. DOI: 10.1039/c4ra16959j

[5] Bafana A, Devi SS, Chakrabarti T. Azo dyes: Past, present and the future. Environmental Reviews. 2011;**19**:350-370. DOI: 10.1139/A11-018

[6] Huang W, Xu J, Lu D, Denga J, Shic G, Zhou T. Rational design of magnetic infinite coordination polymer core-shell nanoparticles as recyclable adsorbents for selective removal of anionic dyes from colored wastewater. Applied Surface Science. 2018;**462**:453-465. DOI: 10.1016/j.apsusc.2018.08.122

[7] Yu X, Liua H, Diaoa J, Suna Y, Wang Y. Magnetic molecularly imprinted polymer nanoparticles for separating aromatic amines from azo dyes—Synthesis, characterization and application. Separation and Purification Technology. 2018;**204**:213-219. DOI: 10.1016/j.seppur.2018.04.081

[8] Deng H, Wei Z, Wang XN. Enhanced adsorption of active brilliant red X-3B dye on chitosan molecularly imprinted polymer functionalized with Ti(IV) as Lewis acid. Carbohydrate Polymers. 2017;**157**:1190-1197. DOI: 10.1016/j.carbpo.2016.10.087

[9] Al-Degs YS, Abu-Surrah AS, Ibrahim KA. Preparation of highly selective solid-phase extractants for Cibacron reactive dyes using molecularly imprinted polymers. Analytical and Bioanalytical Chemistry. 2009;**393**:1055-1062. DOI: 10.10077s00216-008-2502-1

[10] Quinto ML, Khan S, Picassoa G, Sotomayor MPT. Synthesis, characterization, and evaluation of a selective molecularly imprinted polymer for quantification of the textile dye acid violet 19 in real water samples. Journal of Hazardous Materials. DOI: 10.1016/j.jhazmat.2019.121374. in press

[11] Kyzas GZ, Bikiaris ND, Lazaridis NK. Selective separation of basic and reactive dyes by molecularly imprinted polymers (MIPs). Chemical Engineering Journal. 2009;**149**:263-272. DOI: 10.1016/j.cej.2008.11.002

[12] Yan H, Row KH. Characteristic and synthetic approach of molecularly imprinted polymer. International Journal of Molecular Sciences. 2006;**7**:155-178. DOI: 10.3390/i7050155

[13] Chen L, Xua S, Li J. Recent advances in molecular imprinting technology: Current status, challenges and highlighted applications. Chemical Society Reviews. 2011;**40**:2922-2942. DOI: 10.1039/c0cs00084a

[14] Takeuchia T, Haginaka J. Separation and sensing based on molecular

recognition using molecularly imprinted polymers. Journal of Chromatography B. 1999;**728**:1-20. DOI: 10.1016/s0378-4347(99)00057-2

[15] Pichon V, Chapuis-Hugon F. Role of molecularly imprinted polymers for selective determination of environmental pollutants—A review. Analytica Chimica Acta. 2008;**622**: 48-61. DOI: 10.1016/j.aca.2008.05.057

[16] Guć M, Schroeder G. The molecularly imprinted polymers influence of monomers on the properties of polymers—A review. World Journal of Research and Review. 2017;**5**(6):36-47

[17] Panagiotopoulou M, Beyazit S, Nestora S, Haupt K, Tse Sum Bui B. Initiator-free synthesis of molecularly imprinted polymers by polymerization of self-initiated monomers. Polymer. 2015;**66**:43-51. DOI: 10.1016/j.polymer.2015.04.012

[18] Zhenga X, Zhenga H, Xiong Z, Zhao R, Liu Y, Zhao C, et al. Novel anionic polyacrylamide-modify-chitosan magnetic composite nanoparticles with excellent adsorption capacity for cationic dyes and pH independent adsorption capability for metal ions. Chemical Engineering Journal. DOI: 10.1016/j.cej.2019.123706. in press

[19] Luo X, Zhana Y, Huang Y, Yang L, Tu X, Luo S. Removal of water-soluble acid dyes from water environment using a novel magnetic molecularly imprinted polymer. Journal of Hazardous Materials. 2011;**187**:274-282. DOI: 10.1016/j.jhazmat.2011.01.009

[20] Khan S, Wong A, Zanoni AVB, Sotomayor MPT. Electrochemical sensors based on biomimetic magnetic molecularly imprinted polymer for selective quantification of methyl green in environmental samples. Materials Science & Engineering C. 2019;**103**:109825-109832. DOI: 10.1016/j.msec.2019.109825

[21] Foguel MV, Pedro NTB, Wong A, Khan S, Zanoni MVB, Sotomayor MPT. Synthesis and evaluation of a molecularly imprinted polymer for selective adsorption and quantification of acid green 16 textile dye in water samples. Talanta. 2017;**170**:244-251. DOI: 10.1016/j.talanta.2017.04.013

[22] Okutucu B, Sanlıer SH. Decolorization of textile wastewater by dye-imprinted polymer. Desalination and Water Treatment. 2016;**57**:21577-21584. DOI: 10.1080/19443994.2015.1123652

[23] Hu X, Hu Y, Xu G, Lia M, Zhu Y, Jiang L, et al. Green synthesis of a magnetic β-cyclodextrin polymer for rapid removal of organic micro-pollutants and heavy metals from dyeing wastewater. Environmental Research. 2020;**180**:108796-108807. DOI: 10.1016/j.enves.2019.108796

[24] Wang S, Peng Y. Natural zeolites as effective adsorbents in water and wastewater treatment. Chemical Engineering Journal. 2010;**156**:11-24. DOI: 10.1016/j.cej.2009.10.029

[25] Reeve PJ, Fallowfield HJ. Natural and surfactant modified zeolites: A review of their applications for water remediation with a focus on surfactant desorption and toxicity towards microorganisms. Journal of Environmental Management. 2018;**205**:253-261. DOI: 10.1016/j.envman2017.09.077

Chapter 10

Considerations Regarding the Research for the Conservation of Heritage Textiles in Romania

Ilieș Dorina Camelia, Herman Grigore Vasile, Caciora Tudor, Ilieș Alexandru, Indrie Liliana, Wendt Jan, Axinte Anamaria, Diombera Mamadou, Lite Cristina, Berdenov Zharas and Albu Adina

Abstract

Textiles are valuable elements that make up Romania's cultural heritage, being unique through the production techniques, materials used and their significance for the Romanian population. Heritage textiles represent bridges between past and present, kept in collections from different types of buildings. Many of them are preserved and exposed in heritage buildings that are open for public viewing and do not benefit from internal microclimate monitoring systems. These things can have serious repercussions on the integrity and conservation status of these fragile materials. The chapter proposes to analyze the approaches used in different studies to evaluate the risks to which the historical textile collections from Romania are exposed, depending on the place and the way in which they are kept. All these approaches aim to determine the degree of conservation of the materials and their implications on the health of the people with whom they come into contact. Based on the methodology applied in the studies already published, examined in the first part of the chapter, in the second part, a case study was performed on a different sample of historical textiles from Romania. This comes to complete the sphere of knowledge in the field.

Keywords: textiles, heritage buildings, historical textiles, methodological approaches, microclimate, protection, conservation

1. Introduction

The textile materials represented a major coordinated on the evolution of human society both temporally and spatially. The objects made of textile materials have, over time, experienced a great typological, functional and structural diversity, in close relation with the needs of the human society at a certain stage regarding the technological progress achieved and means of processing the textile materials. Therefore, textile objects are part of the cultural heritage, with roles and functions in asserting the specific identity of the place [1–4]. The specificity and identity of the objects made out of textile are derived from the genetic, evolutionary and qualitative features that they incorporate (occupations, customs, technological level achieved, beliefs, superstitions, etc.) [5].

The mutations that have occurred lately against the backdrop of globalization call for sustained efforts to conserve these elements of cultural heritage and local identity. The necessity of their conservation has required the carrying out of numerous specialized studies that have highlighted the anthropogenic and environmental degradation factors [1, 6–8] and their effects on textile, yellowing, decreased elasticity and tear resistance, microflora development [9–15], as well as the necessary measures to prevent their degradation and to conserve them [16–19].

Considering that one of the most objective and important factors with direct effects on the degradation of textile is "time itself", a solution may represent digitization, constituting an informational bridge between past, present and future [20–23]. During the past decades, digitization has become one of the main concerns of cultural institutions and governments all over the world [22] as there is a constant need for preserving cultural heritage in a digital form as well. From a process that transforms a physical object into a digital one [21], digitization became an art itself involving various specialists from different domains having the same aim—preservation. But, still the final goal of digitization is not only to turn a material object into bytes, but making it accessible for everyone interested in it, in an online environment [24] as culture is considered a basic human right [25]. Thus, through digitization the past is connected to the future as the digital world may seem a better option for preserving and conservation of the cultural heritage resources. We are living in the "Digitization Age" and our society understood that cultural heritage belongs to the mass and plays an essential role not only for the social and economic development of the present society but for the future one as well.

From all the cultural resources, maybe the most sensitive in preserving are textile [3] as requires certain storage conditions. Temperature, humidity, light exposure and the mycological content in the composition of the materials are few parameters to be analyzed in order to assure proper storage conditions. In this context, digitization seems a better solution for preservation and promotion, their life span will be undeniable higher although it will be in the cyberspace not in the real world.

Among tests, an analysis that can be performed on textile materials for their digitization, the SEM analysis and digital radiography are highlighted practices whose results have a high degree of plausibility.

The scanning electron microscope (SEM) allows three-dimensional views of external morphology of the fibers, in order to determine their level of conservation [26]. SEM is used to identify fibers and to assess the level and type of fiber wear, degradation or structural alteration [27–29]. The environmental scanning electron microscope (ESEM) can be used to "characterise the surface, interface and dynamic properties of textile materials" [30].

X-radiography and digital radiography uses non-destructive techniques very useful in the study of textile materials, documentation and interpretation in order to conserve valuable objects from cultural heritage [31–33]. The limitation of this method is inaccessibility on a large scale due to the high costs. Another technique for studying old textile materials is the one proposed by Ahmed et al. [34]—laser induced breakdown spectroscopy (LIBS), which also provides chemical information; also for the determination of the chemical composition of the colors used, we mention the use of X-ray fluorescent spectrometry (Spectroscan Max G, Spectron), with the type of spectrometer being wavelength dispersive [35].

Mannes et al. [36] propose a non-invasive approach for the analysis of fragile materials of cultural heritage, neutron and X-ray imaging. Valuable information about the old manufacturing techniques used in the creation of textile but also about the possible treatments for cleaning and preserving them for a longer period of time can be obtained using all the techniques and X-ray spectroscopy methods: scanning electron microscopy with energy dispersive X-ray spectroscopy (SEMEDX), X-ray fluorescence spectroscopy (XRF), particle induced X-ray spectroscopy (PIXE) and also Rutherford backscattering spectroscopy (RBS) [37].

A simple technique that could be encountered for analyzing the colors of the cultural heritage objects is colorimetry (CIELab color system). This technique might be useful when a change in color is desired to be quantified when varying the parameters of the microclimate [38]. Moreover, colorimetry could be a suitable method to determine the change in color when a preservation treatment is tested on a heritage object [39], as the modification of its main features should be avoided.

The color of a cultural heritage object is given by pigments or dyes. These substances can be evaluated from the compositional point of view by using infrared spectroscopy (Fourier Transform infrared spectroscopy). Usually, this technique requires the use of a potassium bromide pellet, which is a time-consuming and destructive operation [40]. Fortunately, a new version of this technique, called attenuated total reflectance FT-IR, was developed, having the main advantage of its non-destructive fashion, as it allows the measurements to be performed without any previous preparation [41]. ATR-FTIR is a valuable technique when analyzing dyes and pigments because it can offer information about the molecular fragments of this substances [42, 43]. In this way, it is possible to determine what type of dye or pigment has been used in the original process of painting. This information could be very helpful when choosing the restoration materials.

But without a doubt, the best way to conserve historical textiles is to keep them in an interior environment that is less harmful for them. The great diversity of the buildings in which they are exposed (museums, traditional constructions, private collections, etc.) determines a great variety of indoor microclimate conditions, which could have repercussions on the degree of conservation of textile collections. Thus, of a major importance is both the determination of the state of conservation of the textiles, and the establishment of the influence that the microclimate from the different storage spaces exerts on them.

In this context, the present study aims to highlight some methodological aspects and the results obtained following their implementation in a series of researches carried out on three structural elements defining for the Romanian society, in which different collections of textiles are kept: wooden church "Saint Martyrs Constantin Brancoveanu and his sons", The Museum House from Sălacea and National Archives of Romania—Bihor County Service from Oradea Municipality (**Table 1**). About the three locations where the research was carried out, we can say that they are representative regarding the Romanian rural communities, offering different environment conditions, including the anthropic impact with implications in manifesting in a form or another the negative effects of the textile objects hosted. Nevertheless, in order to diversify and deepening the obtained results by the research team from University of Oradea and collaborators, in the last part of this chapter a new case study regarding a traditional women's shirt over 100 years old from a private collection, has been conducted.

2. Case studies

This paper aims to analyze from the point of view of approach, methodology and research object, three case studies already carries out regarding the current state and the conservation conditions of some textile collections from three buildings in Bihor County, Romania (**Figure 1**) serving different purposes. These three collections have been extensively examined in numerous specialized papers, some being already published and some being in print.

No.	Location of study	*Title of the studies*—Status (journal, volume, issue, page)—Year
1	Wooden church "*Saint Martyrs Constantin Brancoveanu and his sons*"	*Investigations on air quality in the historic wooden church in Oradea city, Romania*—Published (Environmental Engineering and Management Journal, 17, 11, 2731–2739)—2018
2		*Study on microbial and fungal contamination of air and wooden surfaces inside of a historical Church from Romania*—Published (Journal of Environmental Biology, 39, 6, 980–984)—2018
3		*Indoor air quality assessment and its perception. Case study—historic wooden church, Romania*—In print (Romanian Biotechnological Letters)—2020
4		*Exploring the Indoor Environment of Heritage Buildings and its Role in the Conservation of Valuable Objects*—Published (Environmental Engineering and Management Journal, 18, 12, 2579–2586)—2019
5		*Preserving textile objects in Romanian Wooden Churches. Case study of the heritage wooden church from Oradea*—In print (Industria Textila Journal, 2)—2020
6		*Spectrometry Study of Heritage Objects for the Digitisation of Cultural Heritage—In print (Environmental Engineering and Management Journal)—2020*
7	Sălacea Museum House	*Indoor air quality of museums and conservation of textiles art works. Case study: Salacea Museum House, Romania*—Published (Industria Textila Journal, 70, 1, 88–93)—2019
8		*Analyzing indoor museum air quality implications: Case study of Salacea Museum House in Romania*—Conference paper (Global and Regional in Environmental Protection GLOREP, Ed. Politehnica, 89–91)—2018
9		*Microbial and fungal contamination of air and wooden surfaces inside of Museum House Salacea, Romania*—Under evaluation—2020
10	National Archives of Romania—Bihor County Service	*Microclimatic characteristics and air quality inside of the National Archives of the Bihor County (Romania) for the long-term preservation of the documents and the health of the employees*—Under evaluation—2020
11		*SEM investigations on old maps with canvas support*—Conference paper (International Conference TexTeh IX. Advances textiles for a better world, Proceedings, 9, 153–157)—2019

Table 1.
Details of the studies already carried out by the research team from the University of Oradea and collaborators, on the three buildings in which textile collections are kept.

In the first case, it is a wooden church located in the University of Oradea Campus, built in the second half of the eighteenth century and currently considered a historical monument, being registered in the list of historical monuments with the code BH-II-m-B-20958. It houses numerous paintings, some of which are made on textile material (**Figures 2** and **3**). These are very valuable in terms of the techniques used for painting (al secco), the colors used (some with gold composition), the dimensions and materials on which they are made. The studies envisaged for the methodological and result analysis are focused on determining the interior microclimate of the monument [7, 44, 45] as well as the preservation of the valuable objects inside [35, 46, 47].

The second case study is represented by a former nineteenth century peasant household from the village of Sălacea, which was later refurbished and is currently functioning as a village museum. Numerous ethnographic items are exposed inside such as: furniture, traditional fabrics and other household items (**Figure 4**);

precious due to the prominent place formerly occupied in the life of the Romanian village, especially in the villages from Bihor County. The museum house has been the object of study for three scientific works, one focused on the determination and analysis of the internal microclimate [48], and the second on its influence on the conservation of the exposed textile materials [2].

Figure 1.
The location of the three case studies at the Bihor County level.

Figure 2.
Painting on canvas inside the wooden church historical monument depicting The Holy Trinity.

Figure 3.
Painting on canvas from the ceiling of pronaos inside the wooden church depicting the Virgin Mary with Baby Jesus.

Figure 4.
Traditional port and household items from Sălacea Museum House made of textile materials: (1) elements of the traditional port from the Ier Valley; (2) and (3) items for daily use made from different textile materials.

In the last case study, the conditions and the degree of conservation of the documents and textile materials deposited in the National Archives of Romania—Bihor County Service, based in Oradea Municipality—were examined. The researches focused on a thorough analysis of both the main parameters of the internal microclimate from several deposits [49] and of old maps dated between 1895 and 1910 with the support of textile materials (especially cloth) [50].

Due to the different age of the three buildings, the different construction materials and the purpose they serve, the collections of textiles inside them are subjected to various anthropogenic and environmental pressures. If the building of the National Archives of Romania—Bihor County Service is recently built, using modern building materials and access is restricted, so the textiles inside are subject to minimal external influences in order to conserve them for as long as possible; at the opposite pole is the wooden church from the University of Oradea Campus. It is about 260 years old, built mostly of organic material (wood) and is a "living" monument (it still serves the purpose for which it was built) that houses an average of 60 parishioners at each religious service [7], the paintings on the canvas inside are exposed to much amplified pressures. As for the Museum House in Sălacea, it was built in the last century, made of beaten clay and covered with reed; functioning as a village museum, it is occasionally visited by tourist groups, the textiles not being exposed to a very high stress (at least anthropic).

2.1 The methodology used and its relevance

The three case studies already conducted are based on a common principle regarding methodology; namely the monitoring of the internal microclimate of the spaces where the textiles are stored in order to determine the characteristics of its main elements (temperature, humidity, luminosity, CO_2, contamination with bacteriological microflora and fungi). This has been achieved because it is known that historical textiles are highly susceptible to being damaged by the action of these environmental factors [51]. Theoretically, for textiles to be stored in an environment that is most suitable for storage, the indoor temperature should not exceed 22°C, with a relative humidity between 50 and 65% and a brightness between 50 and 80 luces [52]. To determine the temperature, humidity, brightness and CO_2 concentration, a large number of electronic devices were used in the research, most of them being data loggers for the simultaneous monitoring and storage of the values. The values of the temperature and the relative humidity were determined using Klimalogg Pro, Thermal Imaging Camera FLIR I7 and Delta Ohm HD 32.3, brightness with the help of Digi-Sense Data Logging Luxmeter and Luxmeter data logger Extexh SDL400, and the amount of carbon dioxide with Nova 5000.

Regardless of whether the textiles are made from fibers of plant or animal origin, too high temperature can determine tissue weakening and discoloration. The increased relative humidity of the environment can cause its absorption to the textile fibers and as a result of the humidothermal treatments contractions can occur. Mechanical technological processes can result in a loss of elasticity, flexibility and tensile strength [53]. Furthermore, excess natural light can cause the oxidation of polymers from the composition of natural textile fibers, leading, in time, to the breaking of the intermolecular bonds, yellowing of the material, facilitating the breakdown and penetration of microbial enzymes [13], especially in the case of wool.

The pressure that the characteristics of the main parameters of the microclimate exert on the textile materials in general and the historical ones in particular, should not be considered as an intrinsic relation. These factors are interconditioning and acting simultaneously in the process of material degradation. Thus, there is a directly proportional relationship between the amount of natural light and the interior temperature, as well as inversely proportional between the temperature and relative humidity variations. Further, all these environmental factors contribute to the emergence and spread of microorganisms (such as bacteria and fungi), which present an increased risk of deterioration for materials of organic origin, such as textiles. The most favorable conditions for the colonies of fungi and bacteria are high temperatures (between 24 and 30°C) and very high relative humidity [51]. As a consequence of the activity of certain groups of microorganisms on textiles, we can mention: fiber degradation, discoloration, loss of structural resistance, shading of the affected areas in red, brown, orange or black, cracking and fragmentation areas [1]. For the determination of microbial contamination, for all three case studies the method of Koch sedimentation of the conventional techniques of open plates was used, the samples being collected both from the air and from the surfaces of the textiles. The sampling of the types of fungi and bacteria was done by microscope analysis, and the calculation of the average value of the number of colonies of fungi was materialized by applying Omeliansky's method [54].

If all these methods presented above were of a general nature, being used to establish the influence that the internal microclimate exerts on the textile collections of each of the three analyzed buildings, further on an individual investigation of the fabric samples was carried out. In the wooden church, it was analyzed and represented in cartographic form, with the help of GIS techniques [55, 56], the areas of paintings on canvas anthropic degraded, as well as the distribution of temperatures and brightness within the frescoes. Finally, the colors used for painting inside were examined from the point of view of the internal composition, by digital techniques (X-ray fluorescent spectrometry) [35]. A digital technique was also used to examine two samples collected from old maps with textile support deposited in the

National Archives of Romania—Bihor County Service. Namely scanning electron microscope (SEM); a technique often used in studies targeting textiles, but with a high efficiency in determining their degree of conservation. The use of SEM aimed to identify morphological aspects and irregularities of the fibers that make up the fabrics, as well as the presence of microorganisms and dust [47].

For centuries, the purpose of textiles has been to serve man. Offering protection against the glazes of nature through clothing and material or building houses (tents of nomadic populations), means of purchasing food (fishing nets and various traps), mobility in the territory (sails of ships), and many other practical uses. This has not changed until today, fabrics still occupying a central place in the everyday life of modern society. The relationship of interdependence created along the time between man and textile makes it impossible to analyze the latter separately from the creative element. Therefore, in the accomplishment of the three case studies, the influence of the internal microclimate on the conservation of textiles, as well as their integrated action (internal microclimate + textiles) on human health, was considered.

Furthermore, it is worth mentioning that all the interventions for determining and monitoring the degree of conservation of textiles have been chosen so that they are non-invasive for materials; ensuring that they are kept in the best conditions.

2.2 Results obtained

2.2.1 Wooden church "Saint Martyrs Constantin Brancoveanu and his sons"

Following the microclimate measurements carried out in two periods, March-April (2016) and October-December (2018) respectively, a fluctuating evolution of the main parameters was found. Between March and April 2016, the average temperature inside was 25.8°C, with an average relative humidity of 38% [7]; none of these indicators are complying with the rules in force. The situation is improving, and the values approach the ideal parameters (according to the GD no. 1546/18.12.2003, see [52]), at the level of 2018, when between October and December, the temperature has an average value of 21.1°C, and the relative humidity of 44.3% [46]. Oscillating quantities have also been recorded in terms of carbon dioxide. During the religious services it exceeds in multiple times the value of 2000 ppm, so that during the periods without human activity it is constantly maintained between 400 and 500 ppm.

Following analyses with X-ray fluorescent spectrometry, it was determined that these oscillations are responsible for modifying several properties of the pigments that make up the paintings, eventually leading to their degradation [35].

In order to establish the damaged areas of the paintings on the canvas, the mapping of the areas in which they were degraded by temperature, humidity and rainwater infiltrated in the painted canvas was performed. Also in this sense, the distribution and influence of heat and light on the painted canvas were analyzed (**Figure 5**), at different times of the day and in different seasons [47].

Regarding the contamination with bacteriological microflora and fungi, following the laboratory analyses and sampling of results, 47 colonies of bacteria and 31 of fungi colonies were identified in the air [45]. Fungi belong to a number of 18 species, the most common being *Aspergillus sp.*, *Alternaria sp.*, *Absidia sp.*, *Penicillium sp.*, as well as *Rhodotorula* and *Candida*; and four types of bacteria were identified (*Staphylococcus, Micrococcus, Bacillus,* and *Actinomyces*) [44]. The large number of microorganisms identified in the air inside the wooden church can represent a potential danger to the health of the parishioners, but at the same time they can colonize the textile materials [47]. The organic components that characterize

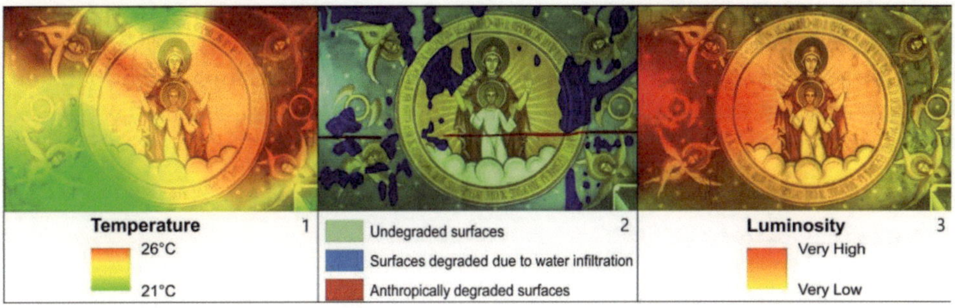

Figure 5.
Distribution of temperature, luminosity and degraded areas on the painting on canvas "Virgin Mary with the Baby Jesus" (after Oana et al., see [42]): (1). Temperature distribution at the canvas level; (2). Distribution of degraded areas at the canvas level; (3). Distribution of natural light at the canvas level.

the paintings on canvas [35] represent a nutritional source for a wide range of microorganisms [45], proving the large number of identified colonies. The samples collected from the paintings on canvas revealed the presence of no less than seven fungi and one type of bacteria [45, 47]. The integrated action of these microbes, both from the air and from the surfaces, can, in time, lead to the decomposition of textile materials, damage that is practically irreversible.

2.2.2 Sălacea Museum House

The measurements made between 03.06.2018–2102.07.2018 in the Museum House in Sălacea Village indicated that the values of temperature, humidity and brightness comply, with small exceptions, the norms set in GD no. 1546/18.12.2003 [2]. Only the temperature recorded an average value higher by 1.3°C than the optimal one (22°C); while humidity, with an average of 65%, is at the upper limit of the ideal. The brightness, having values between 10 and 20 luces, does not influence textiles in any way.

However, these environmental conditions cause the development of microorganisms, which attack the tissues and endanger the health of the visitors. A number of 73 colonies were identified in the center of the main room, 63 colonies in the corners and 39 in the ceiling [2]. Due to the fact that the museum is visited only periodically by tourists, the number of fungal species was obviously smaller than in the wooden church, only three being observed (*Alternaria, Geotrichum* and *Cladosporium*) [48]. However, the action of the bacteriological microflora present inside can cause degradation of textile fibers, as well as health problems (allergies) in humans [6].

2.2.3 National Archives of Romania: Bihor County Service

According to the order No. 235/05.07.1996, supplemented by the Daily Provision of the General Director of the National Archives No. 92/14.05.2009 [57], for the optimum preservation of the documents (including those of textile materials) the microclimate of the interior should be kept between 15 and 25°C regarding the average air temperature, between 40 and 65% the average relative humidity and below 0.3 m/s the speed of air currents [49]. Excluding the relative humidity, which registered a value with 2% (38%) lower than the limit, the other elements are included in these norms (average air temperature of 23.3°C; air currents speed of 0 m/s). The amount of CO_2 recorded an average of 570 ppm, a value considered to be within normal parameters [58].

Aeromicroflora was determined to be composed of fungi species: *Alternaria, Botrytis, Cladosporium, Penicillium, Scopulariopsis* and various subspecies of

Aspergillus [49]. These microorganisms affect human health and integrity of textile fibers; the second case being detected after the SEM analysis of the samples of old maps with textile support. These revealed that the fibers are strongly damaged due to the presence of fungi and dust. Among the microorganisms identified on the maps, there are different subspecies of *Penicillium* and *Fusarium* [50].

3. A new case study

In order to diversify the study objects, a new research on a textile material with historical implications was conducted. It is about a traditional women's shirt called "ie" (**Figure 6**), specific to Alba County, Romania. The value of this fabric lies in its very old age (over 100 years), due to the fact that it is handmade from natural cotton fibers, but also due to the special significance it has in the life of local communities.

Like the other collections of textiles presented above, which are stored in buildings with different environmental conditions, the present study comes to complete the way historical textiles are preserved in Romania. The traditional shirt, being part of a private collection, to which only the owner has access, its preservation conditions are different from all the other samples. It is stored in household microclimate, which commonly consists of an average temperature in the range of 20–24°C (during wintertime) and 23–26°C (during summertime). Also, the humidity level range between 30 and 60% [59]. Moreover, due to the fact that the piece still serves the purpose for which it was achieved, being occasionally worn, it is necessary to determine the bilateral implications of this action; both on the degree of conservation of the shirt and on the health of the wearer.

3.1 Methodology

Based on the methodology applied in previous studies, for the traditional shirt it was decided to evaluate the state of the material, its physical characteristics, the

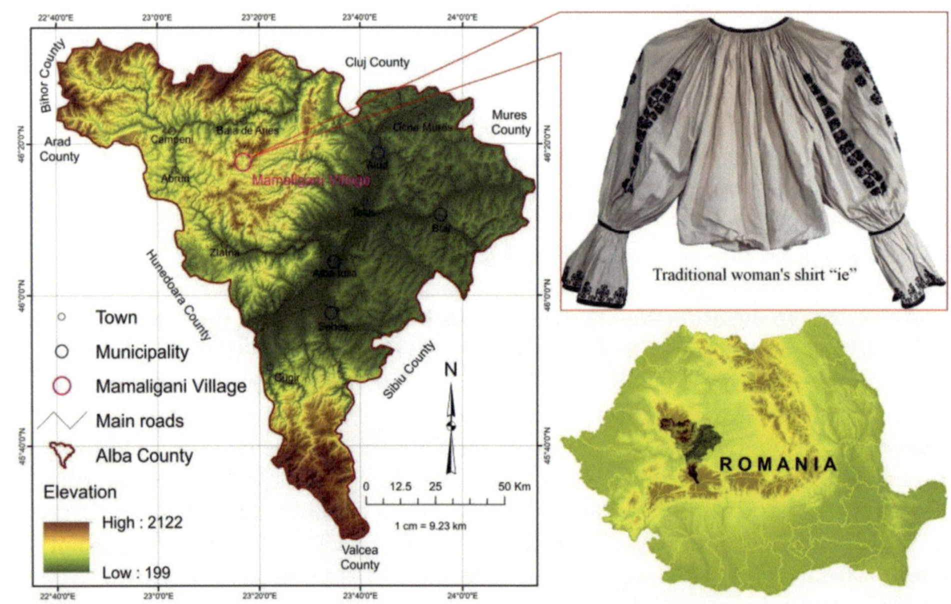

Figure 6.
The geographical location of the area of origin of the traditional women's shirt "ie".

degree of mechanical destruction and the possible influence of atmospheric factors. The samples collected from the textile material were examined using the scanning electron microscope (SEM), to determine the current state of the fibers. The morphology of the fiber system was determined in order to identify visible damage under the microscope and verify the occurrence of dust on the tested impurities. An attempt was made to determine the fungi and other microorganisms potentially occurring on the tested material [60, 61]. The test samples were presented in successively increased magnifications, from 178× to 2.31K× at EHT 20.00 kV. Using the FEI Quanta 200 microscope, a comparative examination was performed between a white and a black thread from the shirt's composition, in order to determine any differences in terms of material or degree of preservation. The whiteness index was determined with the Datacolor Spectrophotometer with a D65/10 lamp, by measuring the X, Y and Z chromatic components from three areas (**Figure 7**) with different bleaching levels.

To identify the microorganisms that are currently colonizing the material, Dichloran culture medium 18% glycerol agar with chloramphenicol (DG 18) was used; the diluent used being peptonated water. Further, petri plates were seeded, and the samples were incubated under aerobic conditions, at 25 ± 1°C, for 5–7 days. Macroscopic and microscopic examination of fungal colonies [62, 63] highlighted the following types of fungi: *Aspergillus niger, Penicillium spp., Cladosporium spp., Alternaria spp. and Candida spp*. After dishes incubation, fungal and bacterial colonies were counted. Colony forming units per cubic meter (CFU • m^{-3}) were determined, following the Omeliansky's equation [64, 65]:

$$N = 5a \times 10^4 \,(bt)^{-1}, \tag{1}$$

where N = microbial CFU • m^{-3} of indoor air, a = number of colonies per Petri dish, b = dish surface (cm^2), t = exposure time (min). Determination of the number of fungi (NTF) per gram of product ufc/g showed NTG/ml or /g = greater than 300 at both 22°C and 37°C, NTF/ml or /g = 2.1×10^3.

3.2 Results and discussions

The first pair of images at 178× magnification (**Figure 8**) and 270× (**Figure 9**) allowed the determination of the tested fiber's condition, as good. The first image

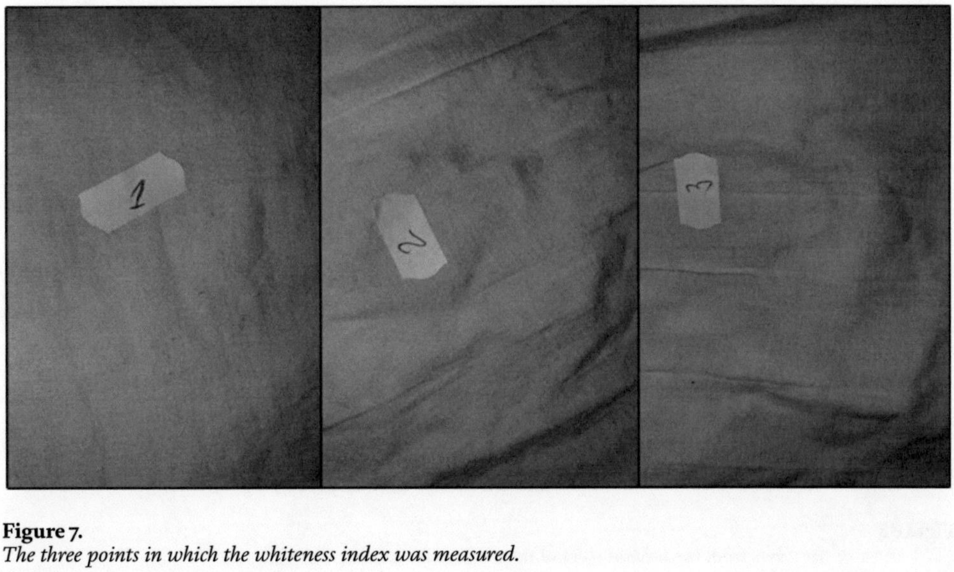

Figure 7.
The three points in which the whiteness index was measured.

clearly shows the weave of threads constituting the basic fiber of the material of the costume under examination. Its good condition may indicate that the folk costume ("ie") is not used very often (within the scope of the test sample), as well as it is kept in good storage conditions. This is confirmed by another image (**Figure 9**), which shows only slight mechanical damage to the fibers. The entire fiber bundle tested is in good condition. However, even at this low magnification, enhanced by image enlargement using computer tools, we can see atmospheric dust covering the fibers.

Successive image magnifications, 818× (**Figure 10**) and 877× (**Figure 11**) confirm previous observation. At this magnification, dust and dirt are clearly visible on individual fibers. Also, it can be observed in **Figure 10**, the structure of a single fiber of natural origin. In the case of natural fibers, with a lose arrangement, the rough, lamellar surface structure means that they are able to transport significant amounts of material, including microbial origin [66]. Computer enlargement of the image of the same fragment of the sample, showing the tested fibers, allows to detect the presence of dust particles and microbiological contamination.

The images of the next sample at 1.31× magnification (**Figure 12**) with clear contamination and at 2.31× magnification with single fiber (**Figure 13**) confirm this observation. The first of them clearly shows mechanical impurities, although the individual fiber in the next image is free of impurities.

Using a datacolor instrument, the whiteness indices of the shirt was measured on three portions, from the portion where the textile is clean to the portions where yellow spots are observed (**Figure 7**). Berger and CIE whiteness indices are presented in **Table 2**. When comparing to literature, the indices for the traditional shirt are significantly lower [67, 68]. This is a proof of the shirt age, even if it has been kept in better conditions. Due to the variations of humidity and temperature, over time the cotton fibers have oxidative reactions, in fact an aging of the cellulose fibers of cotton occurs, thus causing the yellowing of the material.

The SEM images (**Figure 14**) resulting from the analyzes on the two different color yarns (black and white) collected from the shirt, confirm the similarities regarding the cellulose nature of the fibers, both being made of cotton [69]. At the same time, there are no noticeable differences in the state of conservation between the white and black fibers (from the manually embroidered model), both of which are in very good state of conservation.

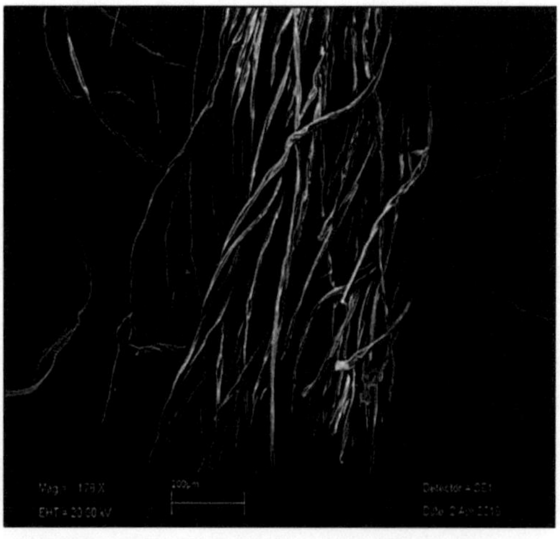

Figure 8.
SEM image of the fibers from traditional shirt at magnification 178×.

DOI: http://dx.doi.org/10.5772/intechopen.91393

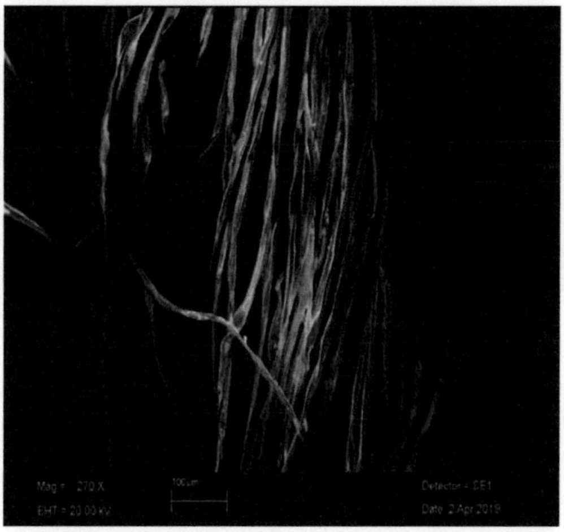

Figure 9.
SEM image of the fibers from traditional shirt at magnification 270×.

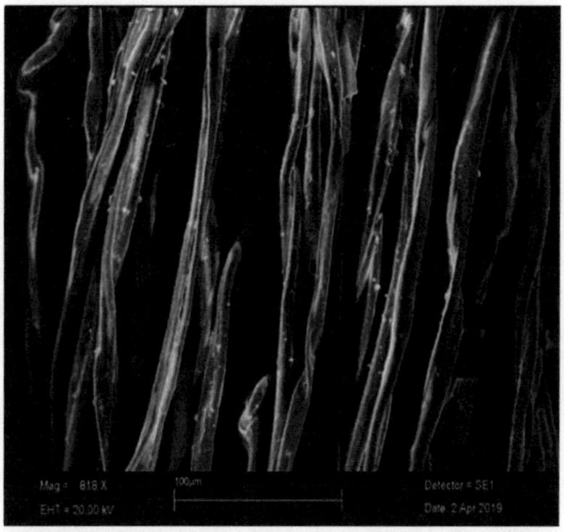

Figure 10.
SEM image of the fibers from traditional shirt at magnification 818×.

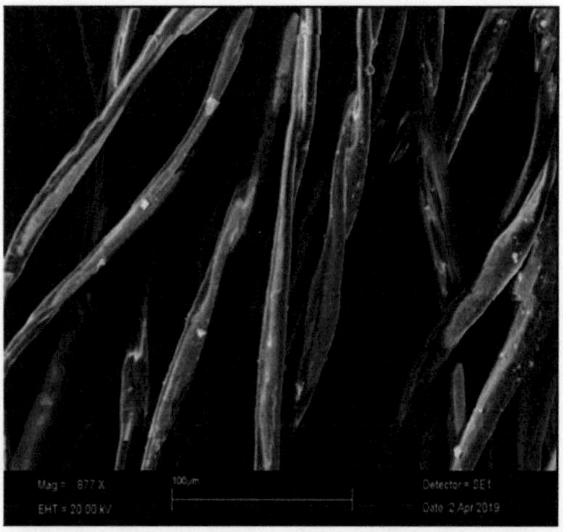

Figure 11.
SEM image of the fibers from traditional shirt at magnification 877×.

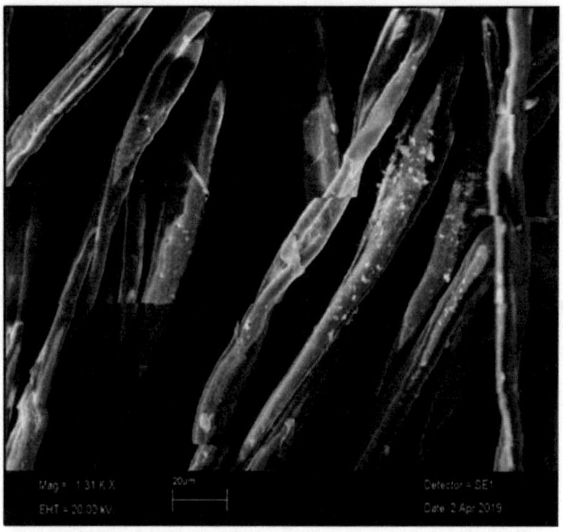

Figure 12.
SEM image of the fibers from traditional shirt at magnification 1.31×.

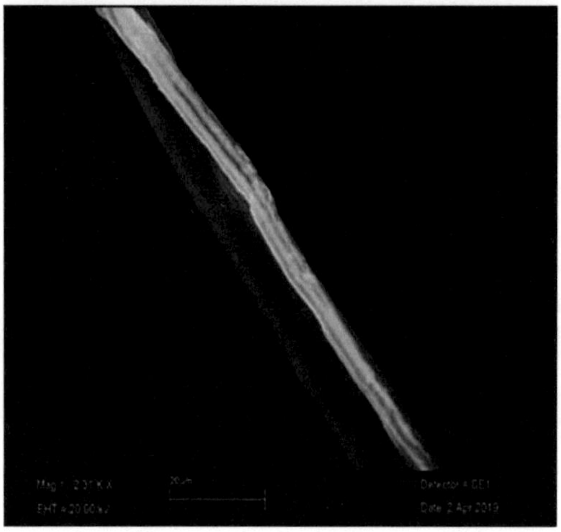

Figure 13.
SEM image of the fibers from traditional shirt at magnification 2.31×.

According to the literature on the subject [66, 70, 71] and the conducted research, it can be stated that the most common fungi found on the tested material (folk costume) and cellulose textiles (cotton, linen) include different species of *Aspergillus, Penicillium, Alternaria, Botrytis, Chaetomium*, etc. [70].

Portion	X	Y	Z	W (Berger)	W (CIE)	T	Obs.
1	65.64	68.92	67.68	46.07	43.86	−3.16	Ref.
2	63.15	66.73	62.74	36.24	30.44	−3.57	Darker More saturated More green
3	63.51	66.69	64.5	41.18	37.34	−3.48	Darker More saturated More green

Table 2.
The whiteness index of the traditional shirt measured in three different portions.

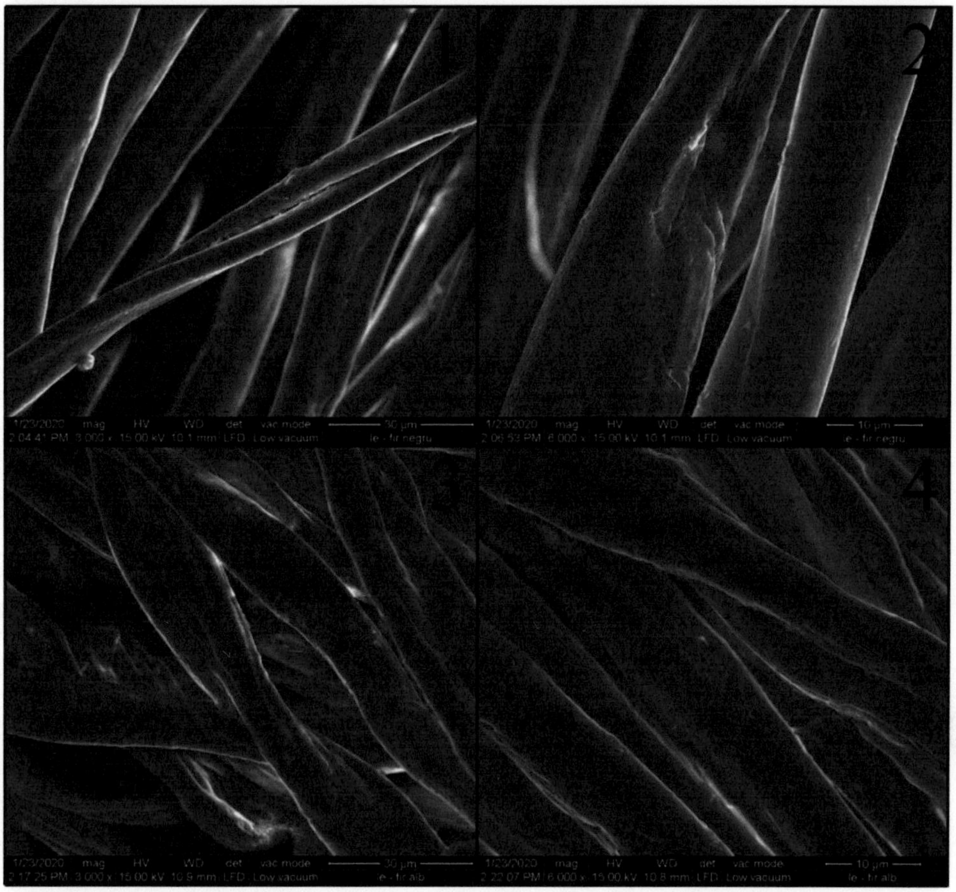

Figure 14.
SEM images of the two white and black wires from the traditional shirt: (1) SEM image of the black wire at 3000× magnification; (2) SEM image of the black wire at 6000× magnification; (3) SEM image of the white wire at 3000× magnification; (4) SEM image of the white wire at 6000× magnification.

The conducted tests of the described fiber samples confirmed the presence of *Aspergillus niger, Penicillium spp., Cladosporium spp. Alternaria spp.* and *Candida spp*. The penultimate enlargement of the tested fabric (**Figure 10**) clearly shows microbiological contamination and the process of fiber development and possible colonization by *Aspergillus* and *Penicillium.*

People are frequently exposed to spores and vegetative forms of *Aspergillus niger*, present on various textile objects. *Aspergillus niger* can cause allergic symptoms and produce certain mycotoxins that can especially affect people with a weak immune system, respiratory, renal, immune system or hearing aids; they can also cause local lesions in both the internal and external ears, postoperative cavities, etc. [72].

Aspergillus and *Candida* cause deaths due to invasive infections. The fungal species in the *Candida* family are the most common etiologic fungal agent of invasive infections that can endanger life in patients: immunocompromised; they have undergone invasive clinical procedures or major trauma and require long-term hospital care [73]. *Cladosporium spp.* can cause allergic reactions in humans, which sometimes results in asthma. Rarely, it can cause opportunistic infections, mainly located in the skin and subcutaneous cellular tissue [74]. Exposure to fungi of the *Alternaria* has been recognized as a risk factor for the development, persistence and severity of asthma and allergic respiratory diseases. They can cause rhinosinusitis, onychomycosis, and skin and subcutaneous infections, generally in immunocompromised patients. Infections with *Penicillium* can be mainly by inhalation and sometimes by ingestion. Pathologies that are the result of infection with *Penicillium spp*. are generally referred to as penicilliosis. Prolonged exposure to fungi from the *Penicillium* family which typically produce mycotoxins ochratoxin A and citrinin, has been reported in combination with opportunistic infections such as keratitis, otomycosis and urinary tract infections [75, 76].

4. Conclusions

In all four case studies, it was proved that the values of the main elements of the microclimate play a decisive role both in the degradation of the textile materials and development in the air and on the surfaces of the bacteriological microflora. In time, its leads to the deterioration of the historical fabrics exposed inside, at the same time endangering human health. The degree of accessibility of the public and the age of the building, is directly proportional to the quality of the elements of the microclimate. Thus, in the wooden church (the oldest building among those analyzed, where human activity is the most intense), textiles are exposed to the highest risk of being degraded due to poor environmental conditions. At the opposite pole, in the Museum House and National Archives of Bihor County, due to the more recent dating of buildings, the nature of the materials and largely limited human activity, the fabrics find better conditions of conservation, and people are exposed to much reduced risks.

By far, in the best state of conservation is the traditional Romanian shirt ("ie"), which due to the fact that it is part of a private collection, is exposed to anthropogenic and lower environmental pressure. It is stored in good conditions, the fibers are very little affected, they do not show discoloration or breakage as a consequence of the action of the microflora.

The analyzes carried out in this study show that some of the historical textiles are in a poor state of conservation, or are threatened to be degraded by human action or environmental factors. In order to conserve the materials for a longer period of time, it is necessary first of all a careful and continuous monitoring of the internal microclimate. The values of its main parameters should be kept as far as possible within the allowed limits, and the impact of the bacteriological microflora and the anthropic factor must be limited. Furthermore, where appropriate, exhibits are preferably to be protected by their installation in special glass boxes to reduce mechanical and chemical damage. All these interventions have both the role of creating a favorable environment for the conservation of textile materials, as well as of minimizing the implications on human health.

Author details

織Ilieș Dorina Camelia[1], Herman Grigore Vasile[1], Caciora Tudor[2*], Ilieș Alexandru[1], Indrie Liliana[7], Wendt Jan[3], Axinte Anamaria[2], Diombera Mamadou[4], Lite Cristina[5], Berdenov Zharas[6] and Albu Adina[7]

1 Department of Geography, Tourism and Territorial Planning, University of Oradea, Oradea, Romania

2 Doctoral School in Geography, University of Oradea, Oradea, Romania

3 Faculty of Oceanography and Geography, Institute of Geography, University of Gdansk, Gdansk, Poland

4 Training and Research Unit of Economic and Social Science, Department of Tourism, University of Ziguinchor Assane Seck, Ziguinchor, Senegal

5 National Research and Development Institute for Textiles and Leather, Bucharest, Romania

6 Department of Physical and Economic Geography, Faculty of Science, Gumilyov Eurasian National University, Nur-Sultan, Republic of Kazakhstan

7 Department of Textiles, Leather and Industrial Management, Faculty of Energy Engineering and Industrial Management, University of Oradea, Oradea, Romania

*Address all correspondence to: tudor.caciora@yahoo.com

References

[1] Gutarowska B, Pietrzak K, Machnowski W, Milczarek JM. Historical textiles—A review of microbial deterioration analysis and disinfection methods. Textile Research Journal. 2016;**87**(19):2388-2406. DOI: 10.1177/0040517516669076

[2] Indrie L, Oana D, Ilieș M, Ilieș DC, Lincu A, Ilieș A, et al. Indoor air quality of museums and conservation of textiles art works. Case study: Salacea Museum House, Romania. Industria Textilă. 2019;**70**(1):88-93. DOI: 10.35530/TT.2019.32

[3] Kumar RA, Sathya RI. Importance of digitization process in textiles. International Journal of Innovative Research in Computer and Communication Engineering. 2015;**3**(7):7011-7015

[4] Petroviciu I, Van den Berghe I, Cretu I, Albu F, Medvedovici A. Identification of natural dyes in historical textiles from Romanian collections by LC-DAD and LC-MS (single stage and tandem MS). Journal of Cultural Heritage. 2012;**13**(1):89-97. DOI: 10.1016/j.culher.2011.05.004

[5] Buchczyk M. To weave or not to weave: Vernacular textiles and historical change in Romania. Textile The Journal of Cloth and Culture. 2014;**12**(3):328-345. DOI: 10.2752/175183514x141563595 36980

[6] Di Carlo E, Chisesi R, Barresi G, Barbaro S, Lombardo G, Rotolo V, et al. Fungi and bacteria in indoor cultural heritage environments: Microbial-related risks for artworks and human health. Environment and Ecology Research. 2016;**4**(5):257-264. DOI: 10.13189/eer.2016.040504

[7] Ilieș DC, Onet A, Marcu F, Gaceu O, Timar A, Baias S, et al. Investigations on air quality in the historic wooden church in Oradea city, Romania. Environmental Engineering and Management Journal. 2018;**17**:2731-2739. DOI: 10.30638/eemj.2018.272

[8] Zavrl SM, Zarnic R. Indoor climate control in building heritage in Slovenia by low temperature wall heating. In: OPET WorkshopLow Exx, Maastricht. Ljubljana, Slovenia: Technological Building and Civil Enginering Institute; 2002

[9] Eryuruk SH, Kalaoğlu F. The effect of weave construction on tear strength of woven fabrics. Autex Research Journal. 2015;**15**(3):207-214. DOI: 10.1515/aut-2015-0004

[10] Gargano M, Rosina E, Monticelli C, Zanelli A, Ludwig N. Characterization of aged textile for archeological shelters through thermal, optical and mechanical tests. Journal of Cultural Heritage. 2017;**26**:36-43. DOI: 10.1016/j.culher.2017.01.009

[11] Yu Y, Cai M, Zhou L, Shao J. Study on the aging and yellowing properties of sebum-soiled cotton fabrics. Fibers and Polymers. 2016;**17**(2):305-309. DOI: 10.1007/s12221-016-5182-z

[12] Kamel FH, Ismael HM, Mohammadamin SA. Microbial deterioration of historical textiles and approaches for their control. International Interdisciplinary Research Journal. 2014;**4**:10-17

[13] Kavkler K, Gunde-Cimerman N, Zalar P, Demšar A. Fungal contamination of textile objects preserved in Slovene museums and religious institutions. International Biodeterioration & Biodegradation. 2015;**97**:51-59. DOI: 10.1016/j.ibiod.2014.09.020

[14] Onet A, Ilies DC, Buhas S, Rahota D, Ilies A, Baias S, et al. Microbial air

contamination in indoor environment of university sport hall. Journal of Environmental Protection and Ecology. 2018;**19**(2):694-703

[15] Triki E, Dolez P, Vu-Khanh T. Tear resistance of woven textiles—Criterion and mechanisms. Composites Part B: Engineering. 2011;**42**(7):1851-1859. DOI: 10.1016/j.compositesb.2011.06.015

[16] Akyuz T, Akyuz S, Balci K, Gulec A. Investigations of historical textiles from the Imperial pavilion (Hunkar Kasri) of the new mosque Eminonu-Istanbul (Turkey) by multiple analytical techniques. Journal of Cultural Heritage. 2017;**25**:180-184. DOI: 10.1016/j.culher.2016.12.009

[17] Quye A. Factors influencing the stability of man-made fibers: A retrospective view for historical textiles. Polymer Degradation and Stability. 2014;**107**:210-218. DOI: 10.1016/j.polymdegradstab.2014.03.002

[18] Patra K. Application of nanotechnology in textile engineering: An overview. Journal of Engineering and Technology Research. 2013;**5**(5):104-111. DOI: 10.5897/jetr2013.0309

[19] Tissera ND, Wijesena RN, Perera JR, de Silva KMN, Amaratunge GAJ. Hydrophobic cotton textile surfaces using an amphiphilic graphene oxide (GO) coating. Applied Surface Science. 2015;**324**:455-463. DOI: 10.1016/j.apsusc.2014.10.148

[20] Comes R, Buna Z, Badiu I. Creation and preservation of digital cultural heritage. Journal of Ancient History and Archaeology. 2014;**1**(2):50-56. DOI: 10.14795/j.v1i2.55

[21] Laine-Zamojska M. Digitization of cultural heritage (Forskningsprojekter). Nordisk Museologi. 2011;**1**(1):125-132

[22] Madanan M, Hussain N, Ahmad KA. A study of the digitization process to preserve the culture and heritage of a civilization using natural language processing and it's impact on the social, economic and scientific aspects. Journal of Theoretical and Applied Information Technology. 2018;**96**(16):5550-5568

[23] Pavlidis G, Koutsoudis A, Arnaoutoglou F, Tsioukas V, Chamzas C. Methods for 3D digitization of cultural heritage. Journal of Cultural Heritage. 2007;**8**(1):93-98. DOI: 10.1016/j.culher.2006.10.007

[24] Sotirova K, Peneva J, Ivanov S, Doneva R, Dobreva M. Digitization of cultural heritage—Standards, institutions, initiatives. In: Ivanova K, Dobreva M, Stanchev P, Totkov G, editors. Access to Digital Cultural Heritage: Innovative Applications of Automated Metadata Generation. Plovdiv: Paisii Hilendarski Plovdiv University Publishing House; 2012. pp. 25-67

[25] Bonici UM. The Human Right to the Cultural Heritage—The Faro Convention's Contribution to the Recognition and Safeguarding of this Human Right. Council of Europe: Heritage and Beyond; 2009. pp. 53-59

[26] Scharff AB, Bolt JL. Evaluating transmission electron microscopy as a method for assessing the condition of archaeological wool. In: Bridgland J, editor. ICOM-CC 18th Triennial Conference; 4-8 September; Copenhagen. Paris: International Council of Museums. 2017. Preprints. art. 1807

[27] Ryde ML, Gabra-Sanders T. The application of microscopy to textile history. Textile History. 1985;**16**(2):123-140

[28] Henry PA. Current techniques of ancient textile analysis: A critical review [Durham E-Theses]. Durham University; 1994. Available from: http://etheses.dur.ac.uk/5599/

[29] Abdel-Kreem O, El-Nagar K. Non-destructive methods to investigate the deterioration of Coptic Egyptian textiles. Journal of Textiles and Apparel Technlogy and Management. 2005;**4**(4):1-15

[30] Wei QF, Wang XQ, Mather RR, Fotheringham AF. New approaches to characterisation of textile materials using environmental scanning electron microscope. Fibres and Textiles in Eastern Europe. 2004;**12**(2):79-83

[31] Garside P, O'Connor S. Assessing the risks of radiographing culturally significant textiles. E-Preservation Science. 2007;**4**:1-7

[32] Zemlicka J, Jakubek J, Kroupa M, Hradil D, Hradilova J, Mislerova H. Analysis of painted arts by energy sensitive radiographic techniques with the pixel detector Timepix. Journal of Instrumentation. 2011;**6**(01):C01066-C01066. DOI: 10.1088/1748-0221/6/01/c01066

[33] Brooks MM, O'Connor SA. X-Radiography of Textiles, Dress and Related Objects. Oxford, United Kingdom: Elsevier/Butterworth-Heinemann; 2007. p. 360

[34] Ahmed HE, Liu Y, Bousquet B, Baudelet M, Richardson M. Investigation of historical Egyptian textile using laser-induced breakdown spectroscopy (LIBS)—A case study. Journal of Textile and Apparel, Technology and Management. 2013;**8**(2)

[35] Demenchuk E, Ilieș DC, Wendt J, Koroleva Y, Ilieș A, Goikhman A, et al. Spectroscopy study of heritage objects for the digitisation of cultural heritage. Under evaluation. Environmental Engineering and Management Journal. 2020

[36] Mannes D, Schmid F, Frey J, Schmidt-Ott K, Lehmann E. Combined neutron and X-ray imaging for non-invasive investigations of cultural heritage objects. Physics Procedia. 2015;**69**:653-660. DOI: 10.1016/j.phpro2015.07.092

[37] Šimić K, Zamboni I, Fazinić S, Mudronja D, Sović L, Gouasmia S, et al. Comparative analysis of textile metal threads from liturgical vestments and folk costumes in Croatia. Nuclear Instruments and Methods in Physics Research Section B: Beam Interactions with Materials and Atoms. 2018;**417**:115-120. DOI: 10.1016/j.nimb.2017.06.015

[38] Abdel-Kareem O. Fungal deterioration of historical textiles and approaches for their control in Egypt. e-Preservation Science. 2010;**7**:40-47

[39] Chadeau E, Oulahal N, Dubost L, Favergeon F, Degraeve P. Anti-*Listeria innocua* activity of silver functionalised textile prepared with plasma technology. Food Control. 2010;**21**:505-512. DOI: 10.1016/j.foodcont.2009.07.013

[40] Abdel-Kareem O. The long-term effect of selected conservation materials used in the treatment of museum artefacts on some properties of textiles. Polymer Degradation and Stability. 2005;**87**:121-130. DOI: 10.1016/j.polymdegradstab.2004.07.014

[41] Smith MJ, Holmes-Smith AS, Lennard F. Development of non-destructive methodology using ATR-FTIR with PCA to differentiate between historical Pacific barkcloth. Journal of Cultural Heritage. 2019;**39**:32-41. DOI: 10.1016/j.culher.2019.03.006

[42] Ahmed HE, Darwish SS. Effect of museum conditions on historical dyed silk fabric with madder dye. Journal of Polymers and the Environment. 2012;**20**:596-606. DOI: 10.1007/s10924-012-0421-x

[43] Cerempei A, Ciobanu L, Muresan E, Măluțan C, Butnaru R. Textile materials functionalised with natural biologically active compounds.

Romanian Biotechnological Letters. 2010;**15**(5):5537-5544

[44] Oneț A, Ilieș DC, Ilieș A, Herman GV, Burtă L, Marcu F, et al. Indoor air quality assessment and its perception. Case study—Historic wooden church, Romania. Romanian Biotechnological Letters. DOI: 10.26327/RBL2019.264

[45] Ilieș DC, Oneț A, Wendt JA, Ilieș M, Timar A, Ilieș A, et al. Study on microbial and fungal contamination of air and wooden surfaces inside of a historical church from Romania. Journal of Environmental Biology. 2018;**39**:980-984. DOI: 10.30892/gtg.26305-393

[46] Ilieș DC, Oneț A, Herman G, Indrie L, Ilieș A, Burtă L, et al. Exploring the indoor environment of heritage buildings and its role in the conservation of valuable objects. Environmental Engineering and Management Journal. 2019;**18**(12): 2579-2586

[47] Oana I, Indrie L, Oneț A, Oana D, Ilieș DC, Hermna GV, et al. Preserving textile objects in Romanian wooden churches. Case study of the heritage wooden church from Oradea, Romania. Industria Textila Journal. 2020;**2**. In print

[48] Ilieș M, Ilieș DC, Oneț A, Baias Ș, Ilieș A, Herman GV, et al. Analysing indoor museum air quality implications: Case study of Salacea Museum House in Romania. In: Timișoara, editor. Global and Regional in Environmental Protection GLOREP; 15-17 November; Romania: Politehnica. 2018. pp. 89-91

[49] Ilieș DC, Gaceu O, Georgiță M, Ilieș A, Baias Ș, Indrie L, et al. Microclimatic characteristics and air quality inside of the National Archives of the Bihor County (Romania) for the long-term preservation of the documents and the health of the employees. Under evaluation. 2020

[50] Ilieș DC, Indrie L, Wendt J, Ilieș A, Gaceu O, Demenchuk E, et al. SEM investigations on old maps with canvas support. In: International Conference TexTeh IX "Advanced Textiles for a Better World"; 24-25 October 2019; Bucharest. Romania: Proceedings. Vol. 9. 2019. pp. 153-157. ISSN 2068-9101

[51] Abdel-Kareem O. Monitoring, controlling and prevention of the fungal deterioration of textile artifacts in the museum of Jordanian heritage. Mediterranean Archaeology and Archaeometry. 2010;**10**(2):85-96

[52] Government Decision of Romania no. 1546 from December 18, 2003 for approval rules for the conservation and restoration of movable cultural assets ranked. Available from: http://www.bcu-iasi.ro/docs/HG1546-2003.pdf

[53] The LS. Textile conservator's manual. In: Butterworth-Heinemann, editor. Conservation and Museology Series. 1998. pp. 17-19

[54] Hayleeyesus SF, Manaye AM. Microbiological quality of indoor air in university libraries. Asian Pacific Journal of Tropical Biomedicine. 2014;**4**:S312-S317. DOI: 10.12980/apjtb.4.2014c807

[55] Lezzerini M, Antonelli F, Columbu S, Gadducci R, Marradi A, Miriello D, et al. Cultural heritage documentation and conservation: Three-dimensional (3D) laser scanning and geographical information system (GIS) techniques for thematic mapping of facade stonework of St. Nicholas church (Pisa, Italy). International Journal of Architectural Heritage. 2014;**10**(1):9-19. DOI: 10.1080/15583058.2014.924605

[56] Campanaro DM, Landeschi G, Dell'Unto N, Leander Touati AM. 3D GIS for cultural heritage restoration: A "white box" workflow. Journal of Cultural Heritage. 2016;**18**:321-332. DOI: 10.1016/j.culher.2015.09.006

[57] The Unit Order of the General Director of the National Archives of Romania no. 235/05.07.1996 supplemented with the Daily Provision of the General Director of the National Archives no. 92 from 14.05.2009. Available from: http://arhivelenationale.ro/site/download/acte_normative/Normativ-depozite-de-arhiva.pdf

[58] ANSI/ASHRAE Standard. Ventilation for Acceptable Indoor Air Quality. Atlanta, GA: American Society of Heating, Refrigerating and Air-Conditioning Engineers, Inc; 2003. Available from: https://www.ashrae.org/file%20library/doclib/public/200418145036_347.pdf

[59] Dan D, Tanasa C, Stoian V, Brata S, Nagy Gyorgy T, Florut SC. Passive house design—An efficient solution for residential buildings in Romania. Energy for Sustainable Development. 2016;**32**:99-109. DOI: 10/1016/j.esd2016.03.007

[60] Amin E. Conservation of a gauze textile from the Tulunid period in Egyptian textile museum, Cairo. International Journal of Conservation Science. 2018;**9**(1):13-24

[61] Amin E. Technical investigation and conservation of a tapestry textile from the Egyptian textile museum, Cairo. Scientific Culture. 2018;**4**(3):35-46

[62] Coman I, Mareș M. In: Junimea, editor. Micologie medicală aplicată. Iasi, Romania: Junimea Publishing House; 2000

[63] Coman I, Gogu M. Ghid Practic de MicologiePerformantica. In: Performantica, editor. Iasi, Romania: Performantica Publishing House; 2009

[64] Bogomolova E, Kirtsideli I. Airborne fungi in four stations of the St. Petersburg underground railway system. International Biodeterioration & Biodegradation. 2009;**63**(2):156-160. DOI: 10.1016/j.ibiod.2008.05.008

[65] Borrego S, Guiamet P, Gómez de Saravia S, Batistini P, Garcia M, Lavin P, et al. The quality of air at archives and the biodeterioration of photographs. International Biodeterioration & Biodegradation. 2010;**64**(2):139-145. DOI: 10.1016/j.ibiod.2009.12.005

[66] Górny R, Ławniczek-Walczyk A, Stobnicka A, Gołofit-Szymczak M, Cyprowski M. Fibers as carriers of microbial particles. Medycyna Pracy. 2015;**66**(4):511-523. DOI: 10.13075/mp.5893.00231

[67] Steen D, Dupont D. Control of structured white textiles. Coloration Technology. 2003;**119**(4):205-211. DOI: 10.1111/j.1478-4408.2003.tb00173.x

[68] Dinu M, Coman D. An experimental study on the improvement of functionality of knitted composite for mattress. In: Bondrea I, Cofaru NF, Inţă M, editors. MATEC Web Conf. 2019. pp. 290-09001. DOI: 10.1051/matecconf/201929009001

[69] Zhao H, Kwak J, Conradzhang Z, Brown H, Arey B, Holladay J. Studying cellulose fiber structure by SEM, XRD, NMR and acid hydrolysis. Carbohydrate Polymers. 2007;**68**(2):235-241. DOI: 10.1016/j.carbpol.2006.12.013

[70] Sterflinger K. Fungi: Their role in deterioration of cultural heritage. Fungal Biology Reviews. 2010;**24**(1-2):47-55. DOI: 10.1016/j.fbr.2010.03.003

[71] Mansour E, Loxton C, Elias RM, Ormondroyd GA. Assessment of health implications related to processing and use of natural wool insulation products. Environment International. 2014;**73**:402-412. DOI: 10.1016/j.envint.2014.08.004

[72] Borrego S, Valdés O, Vivar I, Lavin P, Guiamet P, Battistoni P, et al.

Essential oils of plants as biocides against microorganisms isolated from Cuban and Argentine documentary heritage. ISRN Microbiology. 2012;**1**:1-7. DOI: 10.5402/2012/826786

[73] Brown GD, Denning DW, Gow NAR, Levitz SM, Netea MG, White TC. Hidden killers: Human fungal infections. Science Translational Medicine. 2012;**4**(165:165rv13). DOI: 10.1126/scitranslmed.3004404

[74] Ogórek R, Lejman A, Pusz W, Miłuch A, Miodyńska P. Characteristics and taxonomy of Cladosporium fungi. Mikologia Lekarska. 2012;**19**(2):80-85

[75] Gautam A, Sushil S, Shubhi A, Rekha B. Diversity, pathogenicity and toxicology of *A. niger*: An important spoilage fungi. Research Journal of Microbiology. 2011;**6**:270-280. DOI: 10.3923/jm.2011.270.280

[76] Egbuta M, Mwanza M, Babalola O. Health risks associated with exposure to filamentous fungi. International Journal of Environmental Research and Public Health. 2017;**14**(7):719. DOI: 10.3390/ijerph14070719